Semiconductor Strain Metrology: Principles and Applications

Terence K.S. Wong

Division of Microelectronics
School of Electrical and Electronic Engineering
Nanyang Technological University

eBooks End User License Agreement

DEDICATION

Dedicated to my parents, Jasmine and all my former graduate students in microelectronics

CONTENTS

FOREWORD

The global semiconductor industry uses highly complex processes to manufacture state of the art integrated circuits and microelectromechanical systems for a wide variety of applications. Nowadays, these applications go well beyond the traditional information processing and storage. The measurement of the material and device level properties of a semiconductor is essential knowledge to any practicing engineer and researcher. Due to the dynamic nature of this industry, a large body of measurement techniques had been developed over the past five decades. There are excellent texts such as D.K. Schroder's textbook on the characterization field. However, these texts do not cover the increasingly important topic of strain measurement in semiconductors to any sufficient extent. This eBook is the first attempt to fill this gap in the literature. It is a bridge between the review articles in journals and full length research monographs on one specific topic. The author has made a comprehensive survey of the available strain measurement techniques which are grouped according to the type of electromagnetic radiation or particle beam used to probe the sample. The chapters of this volume follow a uniform structure. For each technique, the essential theory is outlined first with reference to the published literature. This is followed by the experimental method and the capabilities and limitation of the method. Examples of how the method is applied in practice are then given. This format and the conciseness of each chapter will make it convenient to refer and it is not necessary to read the chapters in sequence. This eBook should be a useful addition to the semiconductor literature and its electronic format will ensure it can remain up to date in future.

Rakesh Kumar
Globalfoundries Singapore Pte. Ltd.

PREFACE

The idea of writing this eBook on semiconductor strain metrology occurred to me as a result of supervising a doctoral thesis on strained silicon devices. During the summer of 2005, colleagues from the semiconductor foundry industry expressed to me the need for a simple, in-line measurement technique for the strained silicon devices which were beginning to be introduced into mass production at that time. The quantity to be measured, strain was until then something that is quite extraneous to the device community. As a result, there was no existing technique and equipment that a microelectronics engineer could use to obtain accurate strain data within a silicon field effect transistor. There were of course strain measurement techniques available then but these were developed for bulk materials by mechanical engineers. Hence, they are quite ill suited to the needs of a device engineer which is to find out how strain is distributed within small regions of nanoscale devices. Especially important is how strain is changed by the processing conditions imposed during fabrication. New strain characterization techniques with nanoscale spatial resolution, precision and reproducibility are needed. These must furthermore be compatible with the cleanroom environments of semiconductor manufacturing. They should be in-line and preferably be non-destructive.

Over the course of the next two years, we developed a strain measurement technique that can be called an extension of a standard technique that is used in every semiconductor cleanroom to measure film thickness. The technique is spectroscopic ellipsometry and it was chosen because it requires no new hardware to be procured. Building on the research done on strain effects in semiconductors during the 1970s, we were able to use the strain effect on the variable angle ellipsometric spectra of silicon to deduce the strain in a biaxial strained silicon layer and also obtain agreement with strain deduced by other characterization methods from the same sample. By the time the journal paper for this research was published in Semiconductor Science and Technology in 2007, it was apparent to me that various strain measurement techniques had been developed over the years by researchers from the materials engineering, mechanical engineering and synchrotron radiation communities. Some of these techniques have high spatial resolution and will be of increasing importance in the future. This is because strain effects are also key to the device physics and reliability of nanoelectronic devices such as nanowire transistors, nanophotonic devices and microelectromechanical systems. The literature on these newer strain measurement techniques is very scattered and some of the journals and conference proceedings are not well known to those outside their respective research communities. For this reason, it should be useful to have an up to date (and readily updatable) volume on contemporary strain measurement techniques for the practicing engineer and researchers from different disciplines. The goal of this eBook is to communicate in a concise but quantitative manner the principles and applications of the many different semiconductor strain measurement techniques available.

I thank Bentham Science Publishers for making me aware of this new format of publishing which is especially suitable for the topic of this volume. I would also like to thank the technical staff of the former Chartered semiconductor manufacturing Ltd. for introducing me to this interesting field. Finally, gratitude is expressed to Dr. Grahame C. Rosolen for proof reading several chapters of this volume during the manuscript revision.

Terence K.S. Wong
NTU, Singapore
Email: EKSWONG@ntu.edu.sg

PART 1: INTRODUCTION

CHAPTER 1

Introduction to Strain Metrology for Semiconductors

Abstract Two conventional strain measurement methods, namely the strain gauge and the Moire technique are first discussed. The origin of interest in strain effects in semiconductors is introduced in a chronological sequence beginning with stressed induced defects near isolation structures and the present intentional use of strain in the channel of field effect transistors as a performance booster. The need and the lack of precise strain measurement methods at the submicron and nanoscale are emphasized. Strain measurement for microelectromechanical device materials is also discussed.

Keywords: Strain metrology, Nanoelectronics, Strained silicon, Microelectromechanical systems.

The measurement of the deformation of solid bodies or strain is traditionally the field of interest of civil and mechanical engineers [1]. Strain measurements are important in civil engineering because they provide vital information on the integrity of structures [2]. Mechanical engineers likewise perform strain measurements to monitor the properties of mechanical parts. These strain measurements that are carried out on structural materials typically involve large bulk samples with sizes on the order of centimeters or meters. The measurement methods to be discussed below (briefly) basically involve applying a load to the bulk sample and measuring a change in a property of the sample.

1. PIEZORESISTIVE STRAIN GAUGES

One of the basic strain measurement methods is to use a transducer called the strain gauge to monitor the deformation of a solid [3]. There are two types of strain gauges that are commonly used. In the more common piezoresistive strain gauge (Fig. **1**), the deformation of the gauge material results in a change of the resistance of the strain gauge material and this can be used to deduce the amount of strain undergone by the sample. From basic solid state theory, we know that the resistance R of the semiconductor is given by: $R = \rho l / A$ where ρ is the resistivity; l is the length and A is the cross section area. The resistance change can thus be caused by a change in the length and cross section area of the gauge as well as a change in the resistivity of the gauge material. This can be written mathematically as [3]:

$$\frac{\Delta R}{R} = \frac{\Delta l}{l} - \frac{\Delta A}{A} + \frac{\Delta \rho}{\rho} \tag{1.1}$$

The first term in equation (1.1) is the fractional change in resistance. The second term is the longitudinal strain. The third term is the fractional change in area. It has a minus sign because of the Poisson's ratio to be defined in chapter 2. The last term is related to the piezoresistive effect. The piezoresistive effect was discovered by Smith in 1954 and is the change in resistivity of a metal or semiconductor when it is being deformed [4]. It is also the basic principle behind the strained semiconductor devices that will be discussed later in this chapter. If we take out the strain as a common factor on the right hand side of equation (1.1), then:

$$\frac{\Delta R}{R} = \frac{\Delta l}{l} \left[1 - 2v + P_z \right] \tag{1.2}$$

Here, v is the Poisson's ratio and $P_z = (\Delta \rho / \rho)/(\Delta l / l)$ is the parameter for the piezoresistive effect. The Poisson's ratio will be explained in chapter 2. The term within the square bracket is the sensitivity of a strain gauge or the gauge factor $G = 1 - 2v + P_z$. If the gauge factor G is known from calibration, then the strain can be found from the fractional change in the resistance. In practice, this can be measured readily by using the Wheatstone bridge. Semiconductor materials such as silicon and germanium typically exhibit a larger piezoresistive effect than metals. As a result, the gauge factor of semiconductor gauges is generally higher and thus more sensitive. However, these gauges are also more sensitive to temperature variation which is undesirable. Hence for semiconductor gauges, there is necessarily a tradeoff between strain sensitivity and temperature independence. In practice, highly doped (10^{20} cm^{-3}) p-type semiconductors formed by either ion implantation or diffusion are used as strain gauges. P-type dopants are preferred because they result in greater sensitivity and linearity than n-type dopants. A more complete description of strain gauges can be found in [5, 6].

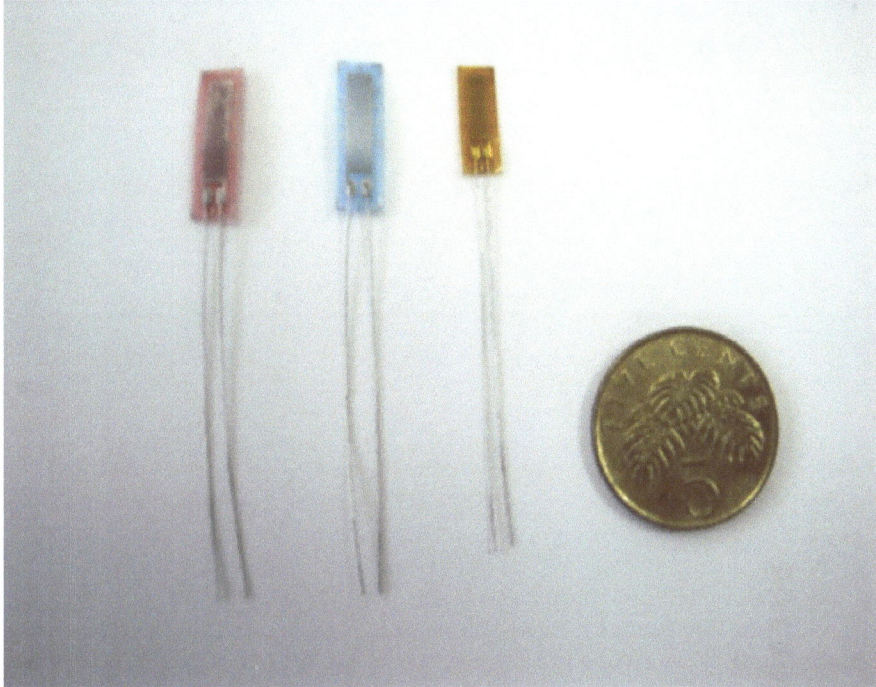

Figure 1: Photograph of piezoresistive semiconductor strain gauges. The strain gauge factor is 2. The diameter of the coin on the right is 16mm.

Figure 2: Moire fringes formed in a digital photograph after scanning.

2. OPTICAL MOIRE METHOD

Another existing strain measurement method is an optical technique known as Moire metrology [7]. The French term 'Moire' was used to refer to a type of textile such as silk that has a granular or wavy appearance. These patterns are sometimes seen when digital photographs are being scanned by a scanner

(Fig. **2**). In the context of metrology, Moire means the optical interference fringes that are formed when two gratings consisting of parallel lines are superposed upon one another. One of these gratings is called the reference grating and it has a constant period p for adjacent lines of the grating. A schematic of a reference grating is shown in Fig. **3** The second grating is called the object grating and is similar except that it is attached to the specimen being measured.

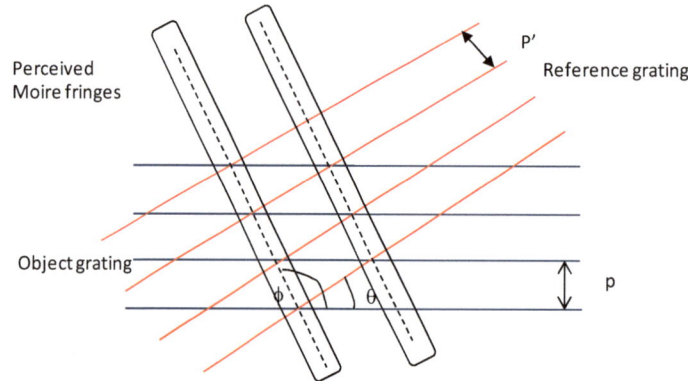

Figure 3: Schematic diagram illustrating the superposition of the reference grating and object grating.

In the following, we will briefly describe the theory of the Moire method because it will be used in chapter 8. The period p' and the angle θ of the object grating with respect to the reference grating can change as a result of the deformation of the specimen. When this occurs, a set of Moire fringes with a distinct period d and orientation angle ϕ can be seen. In Fig. **3**, the brighter bands of the Moire fringes correspond to locations where the lines of the reference and object gratings intersect. The darker bands of the Moire fringes correspond to locations where the two underlying gratings form a zig-zag pattern.

The period d and angle ϕ of the in-plane Moire fringes of Fig. **3** can be derived from geometry to be as follows [7]. The Moire fringe period is given by [7]:

$$d = \frac{pp'}{\left[p^2 \sin^2 \theta + \left(p \cos \theta - p' \right) \right]^{1/2}} \tag{1.3}$$

and the angle is [1.7]:

$$\tan \varphi = \frac{p \sin \theta}{p \cos \theta - p'} \tag{1.4}$$

Note that in both of these equations, the quantity on the left hand side depends on the reference and object grating periods and the angle between these two gratings. For the case of two parallel object and reference gratings, $\theta = 0$ and equation (1.3) will then simplify to [7]:

$$d = \left| \frac{pp'}{p - p'} \right| \tag{1.5}$$

In addition, the Moire fringes will be parallel to the reference grating because of equation (1.4). In the next chapter, we will show that one of the several definitions of linear strain is the Eulerian strain and this is the change in length per unit of final length of the test specimen. In symbols, this can be written as:

$$|\varepsilon| = \left| \frac{l - l_0}{l} \right| = \left| \frac{p - p'}{p'} \right| = \frac{p}{d} \tag{1.6}$$

In this equation, l_0 is the initial length and l is the final length. The second equation in (1.6) can be easily derived from (1.5) and it shows that by measuring the Moire fringe spacing and knowing the reference grating period, the Eulerian strain magnitude can be found. Equation (1.6) also shows the limitations of this method. First, only the magnitude can be deduced because in equation (1.3), we need to take the positive square root of the denominator. This means that the strain value can correspond to either tensile (positive) or compressive (negative) and the sign of the strain will have to be determined by other methods [7]. A second limitation which is common to all the older strain metrology is the lack of spatial resolution. The in-plane Moire fringe method discussed above can only be used to measure uniform strain fields or homogeneous strain. When the strain varies across the specimen, the method will not be able to yield a strain distribution map.

3. STRESS EFFECTS ON SEMICONDUCTOR PROPERTIES

During the 1960s and 1970s, the effect of mechanical stress on semiconductor properties was studied extensively [8-10]. This was a period during which the electronic band structure of many bulk semiconductors were measured experimentally and computed by the application of quantum mechanical principles. Stress was found to be a simple way to modify band structure and to understand the effect of band structure change on semiconductor electronic and optical properties. Important contributors to this field were Manuel Cardona and Fred Pollak of Brown University in the United States and G.L. Bir and G.E. Pikus in the former Soviet Union. The latter pair of researchers wrote an influencial work on symmetry and strain induced effects in semiconductors [11]. Although these early basic research on strain effects were not motivated by technological needs, they are extremely important because they provided the knowledge base for the present strained semiconductor devices. One example is the measurement of deformation potentials in silicon which allowed the amount of band edge shift to be determined.

4. STRAIN MEASUREMENTS FOR NANOELECTRONICS

From this point onwards, this eBook will focus instead on a more recent development in strain metrology, namely the measurement of strain in small volumes within semiconductor thin films that are either deposited or grown epitaxially on planar substrates [12]. These substrates are often a semiconductor material with a different chemical composition. This eBook is also concerned with strain measurement in nanostructures (*e.g.* nanowires) with at least one characteristic length in the range of 1-100nm. The effect of strain on the properties of semiconductors had been studied since the beginning of semiconductor technology in the 1940's. Shockley and Bardeen, co-inventors of the point contact transistor in 1947 studied the effect of strain on the energy bands of semiconductors and developed the deformation potential theory [13]. This theory which is still used today allows the shifts of the conduction and valence band edges to be determined once the strain is known. In the 1950's, Smith working at the Bell Telephone Laboratories carried out the first experimental measurements of strain effects on the electrical properties of germanium and silicon [4]. This effect is nowadays called the piezoresistive effect. The pizeoresistive coefficients of these two elemental semiconductors were measured and the effect was first used for semiconductor strain gauges as mentioned. Currently, it is also the principle behind the strained semiconductor devices that are manufactured commercially [14].

For the semiconductor industry, however, strain effects were regarded more as a reliability concern than as a useful phenomenon. This is because as a result of different thermomechanical properties, many semiconductor processing steps result in residual stress in the silicon wafer. An example is the local oxidation of silicon (LOCOS) process which is a device isolation process that is carried out before transistors are fabricated [15]. In the LOCOS process, a bilayer of silicon oxide (pad oxide) and silicon nitride patterned by optical lithography is used to delineate the active regions of the silicon wafer where the devices will eventually be situated. Then the whole wafer is placed inside an oxidation furnace and the areas of the wafer (isolation regions) not covered by the bilayer is converted to a thick oxide. This is because silicon nitride is a good diffusion barrier. Since a large volume change occurs when silicon is oxidized to silicon oxide, large mechanical stresses often occurred at the periphery of the active regions on the wafer. These stresses can be accentuated by geometric features such as corners of the bilayer. In order to relieve this stress, lattice defects called dislocations will be generated in the silicon. They are undesirable to device performance because they can cause carriers to scatter and decreases the device current.

Another example where strain effects have a negative impact on integrated circuit (IC) performance is the problem of stress migration [16]. This IC failure phenomenon occurs at the conductors or interconnects between transistors and thus affects the semiconductor devices indirectly. Like the silicon substrate, the conductor and insulator layers comprising the interconnect structure of an IC can be under stress after wafer processing. This is because the various materials involved have different thermal coefficient of expansion. These can result in thermal and intrinsic stresses. Again, residual stress needs to be relieved by the interconnect structure. If deformation is not possible because of geometric constraints, stress relief occurs by the formation of voids within the metallic conductors. When their diameter becomes equal to the width of the conductor, an open circuit failure occurs. This can happen even if the circuit has never been powered up. It is clear that local strain measurements are important to metallization as well as semiconductor devices.

By the middle of the 1980s, the role of strain effects in semiconductor technology began to change. Instead of being a cause of reliability issues, strain effects became a useful means for enhancing device performance without changing the chemical composition of the semiconductor [17]. This is due to advances in the field of epitaxy and in particular the growth of silicon-germanium alloys on silicon [18]. Because the lattice constant of germanium is 4% larger than silicon, a silicon-germanium alloy film of any composition is not lattice matched to the substrate. This results in large induced stresses during epitaxy on silicon and only very thin layers with few defects can be grown by an equilibrium method such as liquid phase epitaxy. When thicker layers are attempted, stress relief by formation of dislocations occurs and this resulted in poor quality epitaxial layers. The maximum thickness of epitaxial layer that can be grown without dislocation formation is called the critical layer thickness. It was realized by John Bean and Robert Hull at the Bell Laboratories in 1984 that by using non-equilibrium growth conditions (reduced growth temperature), the critical layer thickness can be increased and high quality pseudomorphic layers of silicon-germanium can be grown by the technique of molecular beam epitaxy (MBE) [14]. In 1992, pseudomorphic layers of silicon-germanium were also grown by ultra high vacuum chemical vapor deposition (UHV-CVD) by Bernard Meyerson at IBM [18]. This process opened the way to the mass manufacturing of heterojunction bipolar transistors for high frequency applications. Actually, the pseudomorphic layers are only meta-stable. Eventually, the layer will in principle revert to the equilibrium state with dislocations. However, due to kinetic limitations, the rate is so slow that the film is for all intents and purposes stable.

The earliest application of pseudomorphic silicon-germanium layers is in the fabrication of the silicon-germanium heterojunction bipolar transistor (HBT) invented by Professor Herbert Kroemer in the 1950s [19]. This device is closely related to the silicon bipolar transistor but has superior properties such as current gain, higher frequency response and larger Early voltage. The reason for its high performance is the presence of a base emitter heterojunction that gives rise to an energy barrier for holes in the valence band. This barrier can be formed by growing a layer of silicon (emitter) on a pseudomorphic layer of silicon-germanium (base) with the same lattice constant as the silicon in the collector. The heterojunction can substantially reduce the base current and brings about the device benefits mentioned above. Since the base layer is typically quite thin, it is relatively straightforward to grow and dope this silicon germanium layer *in situ* by either MBE or UHV-CVD on a collector layer. This is why the HBT was the first commercialized heterojunction device. It should be pointed out that while strain is present in the base layer of a HBT, the carrier transport within the device is not directly affected by the strain. Instead, it is the reduced (and graded) energy band gap due to germanium incorporation in the base that is crucial to device operation.

Starting with a strained silicon-germanium (SiGe) psuedomorphic layer on silicon and by growing additional layers with monotonically increasing germanium content, one can fabricate a top layer of germanium rich silicon-germanium or pure germanium which is fully relaxed (Fig. **4**) [20]. This can be achieved because any dislocations that are generated at the lower part of this graded buffer layer will reduce the stress built up in those lower layers. The relaxed surface layer is often called a metamorphic substrate because its equilibrium lattice constant is different from the supporting wafer and it can be used for further epitaxial growth as if it were an ordinary bulk wafer [20]. If silicon is grown on silicon-germanium or germanium metamorphic substrate, the silicon will become strained because of lattice mismatch. Since the lattice constant of silicon germanium or germanium is larger than silicon, the silicon epitaxial layer will be stretched in the plane of the substrate and it is said to be under biaxial stress. Provided the critical layer thickness for silicon is not exceeded, the silicon layer will not contain lattice defects. This innovation in substrate engineering was first reported by Fitzgerald and colleagues at MIT in 1990 [21]. It opened the way to device performance improvement by what is now called strain engineering.

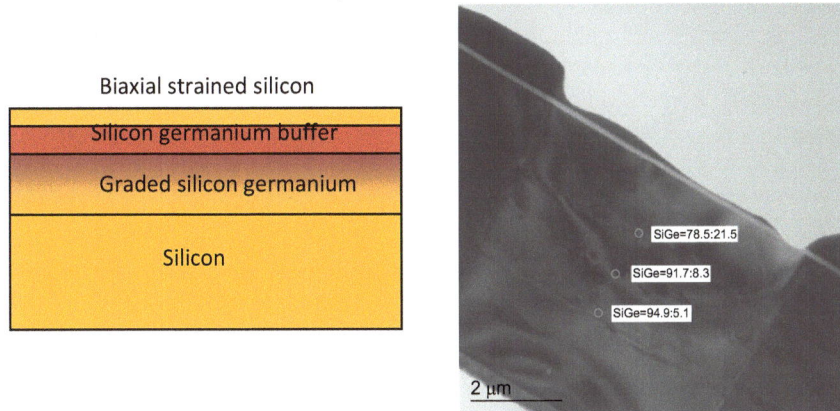

Figure 4: (a) Schematic diagram of a silicon germanium graded buffer layer metamorphic substrate for the growth of biaxial strained silicon; (b) cross sectional TEM image of a biaxial stressed silicon layer on a relaxed silicon germanium metamorphic substrate.

An early success of strain engineering is the use of strained silicon layers to build metal oxide semiconductor field effect transistors (MOSFETs) with enhanced performance [21]. The MOSFET is the most important device to the semiconductor industry and is used in both logic and memory circuits. The traditional approach to improve the MOSFET performance had been to reduce the length of the gate by the scaling approach. This involves applying Dennard's scaling rules and use increasingly high resolution and expensive optical lithography to pattern device structures. By 2004 or the 90nm (half-pitch) technology node in the International technology roadmap (ITRS) for semiconductors, this scaling approach has almost run its course. This means the escalating cost of implementing advanced lithography is making feature size reduction increasingly uneconomical. An alternative approach to device improvement without using simply reduced gate length is needed.

Strained silicon was the first of several performance booster solutions to this problem facing the semiconductor industry [22]. The basic idea is to use the piezoresistive effect or apply stress to a silicon crystal to modify its energy band structure [23, 24]. This is illustrated by appendix 1 which shows two energy momentum *(E-k)* diagram for unstrained silicon and silicon with 0.5% biaxial strain. The *E-k* diagram was computed using the tool Strain Bands in Nanohub. The allowed energy bands along several directions between points of high symmetry in the first Brillouin zone (*Λ, Γ, X* and *K*) were plotted. By comparing appendix 1a and 1b, it is clear that strain has a marked effect on the band structure. If the stress is applied in the correct way, the electron and hole mobilities in n-channel devices can be increased because of reduced inter-valley scattering and reduced effective mass. For p-channel devices, the mobility was found to be enhanced mainly by warpage of the valence band. Mobility enhancement was first demonstrated in n-channel and then p-channel MOSFETs by Hoyt and others in the mid-1990s [21].

Since 2004 or the 90nm technology node of the ITRS, three has been a growing reliance on the use of strain in semiconductor devices by the microelectronics industry [25, 26]. This is because when properly applied, strain can enhance the electrical characteristics of these devices without using prohibitively expensive nanolithography for device patterning. This allows the industry to sustain Moore's law despite the impending end of conventional down scaling of device dimensions. However, what the industry prefers are simple, cost-effective methods with adequate throughput that can extend the performance of conventional MOSFETs without the introduction of costly new process steps. Thus, epitaxial metamorphic substrates or biaxial stress did not find widespread adoption [27]. The stressing schemes used instead are uniaxial or local stress schemes [28, 29]. Typically, the stress is applied only to designated areas of the silicon substrate and the applied stress is generally in the direction of current flow. This stress is induced by an adaptation of standard processing steps such as plasma enhanced chemical vapor deposition (PECVD) (Fig. 5) and selected epitaxial growth [30].

An early example of the uniaxial stress processes is the contact etch stop layer (CESL) process (Fig. 5a) [31]. This process can be integrated right after the self-aligned source drain ion implantation and involves the PECVD of a layer of highly stressed dielectric such as silicon nitride (SiN). Due to its different bonding configuration and thermal properties, the as-deposited SiN is usually highly stressed. This stress can be transferred to the surface region of a silicon substrate with which it is in contact. For n-channel MOSFETs, a tensile stressed SiN layer deposited over the n-MOSFET can enhance device performance. The enhancement is due to a reduction in the amount of inter-valley scattering in the Si conduction band [31].

For p-channel MOSFETs, a compressively stressed SiN layer can boost p-MOSFET performance [30]. There is another uniaxial stressing scheme for p-MOSFETs that is called selective epitaxy of raised silicon germanium source-drain (Fig. 5b) [22]. In this process, the wafer is first deposited with a dielectric and this is then etched through at the source and drain regions of the p-MOSFET. Pseudomorphic SiGe is then grown within the source and drain regions. Since the SiGe has a larger lattice constant than Si, a compressive stress is induced in the silicon channel between the source and the drain. Silicon carbon (SiC) alloys (not to be confused with silicon carbide) can also be grown epitaxially over the source and drain. In this case, the smaller lattice constant of the silicon carbon alloy will result in a tensile stress which will enhance the performance of n-MOSFETs.

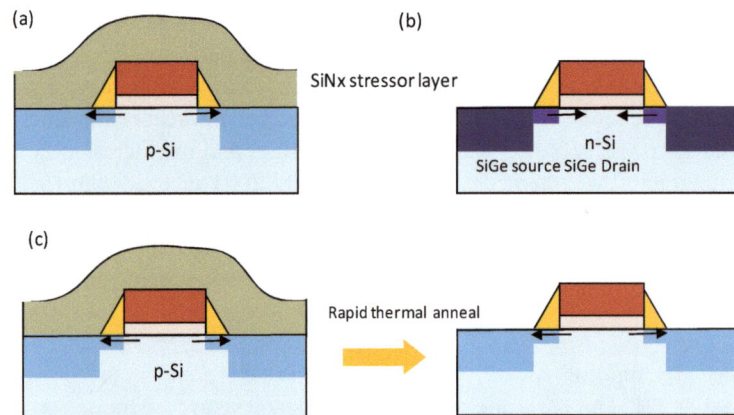

Figure 5: Strained silicon MOSFET stressed by (a) a silicon nitride stressor layer; (b) silicon germanium source drain grown by selective epitaxy and (c) stress memorization technique.

In addition to the above stressor schemes, there is the stress memorization technique (SMT) which is similar to the CESL process mentioned above (Fig. 5c) [32, 33]. The SMT process is found to be effective for n-MOSFETs only and so is not applicable to p-MOSFETs. After the tensile stressed SiN layer is deposited, the wafer is subjected to a brief high temperature anneal called activation. During this step, the stress in the SiN is found to transfer to the Si channel of the n-MOSFET. Thus after the activation anneal, the SiN layer can be removed. For this reason it is also said to be a sacrificial layer.

Finally, an interesting stressing method capable of applying high uniaxial strain in bulk CMOS devices had been developed by Belford and co-workers [34]. The uniaxial methods based on stressor films typically result in strains of the order of 0.1%. However, with the method of Belford [34], strains in the region of 0.1% - 2.5% can be attained. The method is based on thinning a device wafer by lapping and chemical etching until it is 6-10 μm thick. The thinned wafer is then bonded by adhesive to a polymer carrier such as polyimide. Uniaxial longitudinal or transverse stress is applied to the polymer and this stress is transferred to the wafer. This stressing method was used to investigate the current enhancement mechanism in pMOSFETs.

All the uniaxial stressing schemes can enhance the current drive of the devices at a given drain voltage. Today, the strained silicon MOSFET is a standard process offered by all semiconductor device manufacturers. The use of strained silicon is expected to continue into the future when the planar MOSFET architecture is replaced by the multiple gate and silicon nanowire geometries. As with the development of any new manufacturing process, there needs to be concurrent developments in metrology for strained semiconductor devices. New strain measurement techniques are needed to enable device engineers to probe the strain distribution set up by a stressor within a device. This type of information which is lacking at the moment will allow the researcher to tailor the stressor and thus achieve the most effective enhancement of carrier mobility.

In the 2010 update of the International Roadmap for Semiconductors (ITRS) [35], there is a projection of the future developments needed for stress measurements within the front end process metrology section of chapter 9 on metrology. A portion of the accompanying table (MET5) of this section is shown in Table **1** for the period 2009-2016. For off-line stress measurements, manufacturable solutions exist from 2011 until 2016 (yellow boxes). On the other hand, in-line measurements have no manufacturable solutions from 2011 to 2016 and beyond (red boxes). The in-line measurement wafer throughput of 2 wafers per hour for 25 sites per wafer is also lacking at present. Beyond 2017, even the technology targets are not yet determined (as of 2010). Evidently, in the coming technology nodes, new techniques will be needed for characterizing strain at both the transistor level and the micro-area (wafer) level [37]. For in-line applications, any new technique must be non-destructive while off-line applications can be destructive.

Table 1: Stress Metrology Technology Requirements from 2010 Update of ITRS

Metrology for stress/strain in channel and active area	2009	2010	2011	2012	2013	2014	2015	2016
Spatial resolution (nm) of off-line stress measurement at 50MPa resolution	4	3.6	3.2	2.8	2.6	2.2	TBD	TBD
Spatial resolution (nm) of in-line stress measurement at 50MPa resolution	50	45	40	36	32	28	TBD	TBD
Throughput of in-line stress measurement (Wafers/hour at 25 sites/wafer)	2	2	2	2	2	2	TBD	TBD

Accurate strain metrology is also critical to calibrating process simulation tools. In the absence of experimental transistor strain data, the only option at present is to perform computer aided design stress simulations within a device (Fig. **6**). However, the output of this computation is only meaningful if calibrated models are provided by the software used. Thus, strain measurement techniques are critical to the future development of CMOS devices.

5. STRAIN MEASUREMENTS FOR MEMS AND NEMS

Strain measurement of small structures is also of great interest to mechanical engineers specializing in the fields of experimental mechanics, micro and nanoelectromehcanical systems (MEMS and NEMS) [6]. In experimental mechanics, the goal is to understand how structural elements deform and fail under loads. For brittle materials such as semiconductors, a basic question is how does fracture occurs in such materials. Although theories have been developed to explain fracture, it is critical to have experimental data to support these models. Thus measurements of deformation at the vicinity of cracks are extremely useful to improve our basic understanding of mechanical behavior [6].

MEMS is a sub-field of microfabrication technology that evolved from the processing techniques developed for semiconductor microelectronics. In the 1980s, Peterson did pioneering research in bulk micromachining in which wet etching of silicon was used to fabricate micrometer scale mechanical elements such as suspended membranes [6]. This was followed by the development of surface micromachining at the University of California, Berkeley in which photolithography, plasma deposition and etching were used to make mechanical elements such as cantilevers, gears, linear motors and comb drives for transducer and actuator applications [6]. The transducer elements or sensor receives energy (*e.g.* light and heat) from the ambient and converts this energy input into an electrical signal. The actuator performs the reverse operation and converts electrical energy into kinetic energy. Since both sensors and actuators can be fabricated on the same substrate as transistors and interconnects, highly functional and compact integrated microsystem can be made. Some of the most well known MEMS systems include pressure sensors for airbags in vehicles and accelerometers for orientation sensitive portable information displays. For all such MEMS applications, it is necessary to know the mechanical properties of the materials used such as polycrystalline silicon (polysilicon), silicon nitride and silicon dioxide [36]. This is because the mechanical properties of these materials and their behavior at small dimensions can be different from the same material in bulk form. Two of the most basic elastic properties of solid materials are the Young's modulus and the Poisson's ratio. In order to measure both these properties, the strain of the mechanical element needs to be measured. Strain measurement is also needed for the determination of the tensile strength, fracture toughness as well as fatigue and creep behavior. However, as a result of the small dimensions of MEMS structures (of order 10μm or less), it is by no means routine to measure these properties accurately by experiment. This is probably why there is a wide variation in the reported values for the Young's modulus of polysilicon in the literature [6]. In order to address this problem, there had been attempts by the MEMS research community to measure strain at the micrometer length scale. However, the application of these strain measurement to MEMS structures had been very limited.

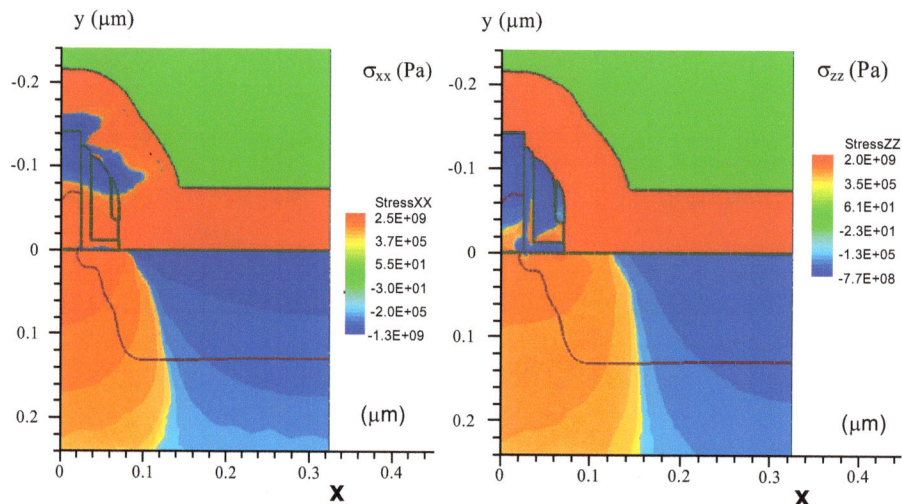

Figure 6: Technology computer aided design (TCAD) simulation of stress distribution within a strained silicon n-channel MOSFET by Sentaurus Process. On the left, the normal stress in the *x* direction is shown as a color coded contour plot. The normal stress in the *y* direction is shown in a similar way on the right.

6. ORGANIZATION OF THE eBook

This first volume of the Metrology Series provides an up to date and comprehensive overview of the rapidly developing interdisciplinary field of semiconductor metrology. It is presented in the form of a handbook and is organized into four parts. Part 2 which follows the next tutorial chapter on strain describes several optical methods in which either visible or ultraviolet photons are used as the probing electromagnetic radiation. The strain measurement methods to be discussed include spectroscopic ellipsometry (chapter 3), photoreflectance method (chapter 4) and micro-Raman spectroscopy (chapter 5). Part 3 describes strain measurement methods that are based on the scanning electron microscope (SEM) and the transmission electron microscope (TEM). The cathodoluminescence method is covered in chapter 6. This is followed by two electron beam diffraction methods in chapter 7 called nanobeam diffraction and convergent beam electron diffraction. The very recent electron holography Moire method is covered in chapter 8. Part 4 groups together a few other specialized strain measurement techniques that had been reported in the recent literature. Chapter 9 discusses the near-field or tip-enhanced Raman spectroscopy. Chapter 10 describes the atomic force microscope and digital image correlation technique. Chapter 11 is concerned with X-ray micro and nanodiffraction techniques in which the sample is probed by an intense focused X-ray beam from a synchrotron source. The reader will realize that several techniques are actually '*in situ*' mechanical measurements of small regions within a sample. This means the sample is placed within a microscope with a sample chamber that is large enough to accommodate both the sample and microfabricated actuators and sensors. These are used to apply stress to the sample and measure the resulting strain in the sample. For ease of reference, each chapter will have the same layout, namely: introduction, measurement principle, experimental technique, resolution and limitations and application examples. The examples are drawn from the latest journal literature and the Frontier of Characterization and Metrology conference series. The latter which had been held regularly since 1999 is the flagship conference of this field. Together the examples serve to illustrate how the methods are used in practice. A list of references is included for each chapter so that further information can be found if needed.

REFERENCES

[1]   P.C. Chou and N.J. Pagano, *Elasticity: Tensor, Dyadic and Engineering Approaches,* Dover: New York, 1975.

[2]   J.E. Gordon, *The Science of Structures and Materials,* Scientific American Publishers: New York, 1988.

[3]   K.K. Ng, *Complete Guide to Semiconductor Devices,* McGraw-Hill: New York, 1995.

[4]   C.S. Smith, "Piezoresistance effect in germanium and silicon," *Phys. Rev.,* vol. 94, pp. 42-49, Apr. 1954.

[5]   S.M. Sze and K.K. Ng, *Physics of Semiconductor Devices,* Wiley Interscience: New York, 2006.

[6]   C. Liu, *Foundation of MEMS,* Prentice Hall: Upper Saddle River, 2006.

[7]   T. Yoshizawa, Ed., *Handbook of Optical Metrology Principles and Applications,* CRC Press: Roca Baton, 2008.

[8]   M. Cardona and F. Pollak, "Piezo-electroreflectance in Ge, GaAs and Si," *Phys. Rev.,* vol. 172, pp. 816-837, Aug. 1966.

[9]   E.O. Kane, "Strain effects on optical critical-point structure in diamond type crystals," *Phys. Rev.,* vol. 178, pp. 1368-1398, Mar. 1969.

[10] F. Pollak, "Modulation spectroscopy under uniaxial stress," *Surf. Sci.,* vol. 37, pp. 863-895, Jun. 1973.

[11] G.L. Bir and G.E. Pikus, *Symmetry and Strain-Induced Effects in Semiconductors,* Wiley: New York, 1974.

[12] P.R. Chidambaram, C. Bowen, S. Chakravarthi, C. Machala and R. Wise, "Fundamentals of silicon material properties for successful explotation of strain engineering in modern CMOS manufacturing," *IEEE Trans. Electr. Dev.,* vol. 53, pp. 944-964, May. 2006.

[13] H.H. Hall, J. Bardeen and G.L. Pearson, "The effect of pressure and temperature on the resistance of p-n junctions in germanium," *Phys. Rev.,* vol. 84, pp. 129-132, Oct. 1951.

[14] C.K. Maiti, *Strained Silicon Heterostructures: Materials and Devices,* IEE: London, 2001.

[15] R.C. Jaeger, *Introduction to Microelectronic Fabrication,* Addison Wesley: Reading, 1993.

[16] S.M. Hu, "Stress related problems in silicon technology," *J. Appl. Phys.,* vol. 70, pp. R53-R80, Sep. 1991.

[17] K. Rim, R. Anderson, D. Boyd, F. Cardona, K. Chan, H. Chen, S. Christiansen, J. Chu, K. Jenkins, T. Kanarsky, S. Koester, B.H. Lee, K. Lee, V. Mazzeo, A. Mocuta, D. Mocuta, P.M. Mooney, P. Oldiges, J. Ott, P. Ronsheim, R. Roy, A. Steegen, M. Yang, H. Zhu, M. Ieong and H.S.P. Wong, "Strained Si CMOS (SS CMOS) technology: opportunities and challenges," *Solid State Electr.,* vol. 47, pp. 1133-1199, Jul. 2003.

[18] J.D. Cressler, *The Silicon Heterostructure Handbook: Materials, Fabrication, Devices, Circuits and Applications of SiGe and Si Strained-Layer Epitaxy,* CRC Press: Roca Baton, 2006.

[19] Y. Taur and T.H. Ning, *Fundamentals of Modern VLSI Devices.* Cambridge University Press: Cambridge, 1998.

[20] E.A. Fitzgerald, "Engineered substrates and their future role in microelectronics," *Mater. Sci. Eng. B,* vol. 124-125, pp. 8-15, Dec. 2005.

[21] M.L. Lee, E.A. Fitzgerald, M.T. Bulsara, M.T. Currie and A. Lochtefeld, "Strained Si, SiGe and Ge channels for high-mobility metal-oxide-semiconductor field-effect transistors," *J. Appl. Phys.,* vol. 97, pp. 011101-1-27, Jan. 2005.

[22] M. Ieong, B. Doris, J. Kedzierski, K. Rim and M. Yang, "Silicon device scaling to the sub-10-nm regime," *Science,* vol. 306, pp. 2057-2060, Dec. 2004.

[23] Y. Sun, S.E. Thompson and T. Nishida, "Physics of strain effects in semiconductors and metal-oxide-semicondutor field-effect transistors," *J. Appl. Phys.,* vol. 101, pp. 104503-1-22, May. 2007.

[24] S.E. Thompson, G. Sun, Y.S. Choi and T. Nishida, "Uniaxial-process-induced strained-Si: extending the CMOS roadmap," *IEEE Trans. Elect. Dev.,* vol. 53, pp. 1010-1020, May. 2006.

[25]   S.E. Thompson, M. Armstrong, C. Auth, S. Cea, R. Chau, G. Glass, T. Hoffman, J. Klaus, Z. Ma, B. Mcintyre, A. Murthy, B. Obradovic, L. Shifren, S. Sivakumar, S. Tyagi, T. Ghani, K. Mistry, M. Bohr and Y. El-Mansy, "A logic nanotechnology featuring strained-silicon," *IEEE Electr. Dev. Lett.,* vol. 25, 191-193, Apr. 2004.

[26]   S.E. Thompson, M. Armstrong, C. Auth, M. Alavi, M. Buehler, R. Chau, S. Cea, T. Ghani, G. Glass, T. Hoffmann, C.H. Jan, C. Kenyon, J. Klaus, K. Kuhn, Z. Ma, B. Mcintyre, K. Mistry, A. Murthy, B. Obradovic, R. Nagisetty, P. Nyguen, S. Sivakumar, R. Shaheed, L. Shifren, B. Tufts, S. Tyagi, M. Bohr and Y. El-Mansy, "A 90-nm logic technology featuring strained-silicon," *IEEE Trans. Electr. Dev.,* vol. 51, pp. 1790-1797, Nov. 2004.

[27]   R. Harper, "Epitaxial engineered solutions for ITRS scaling roadblocks," *Mater. Sci. Eng. B,* vol. 134, pp. 154-158, Oct. 2006.

[28]   J.-G. Park, C.-S. Lee, T.-H. Kim, S.-H. Hong, S.-J. Kim, J.-H. Song and T.-H. Shim, "Strained Si engineering for nanoscale MOSFETs," *Mater. Sci. Eng. B*, vol. 134, pp. 142-153, Oct. 2006.

[29]   P.M. Mooney, "Improved CMOS performance *via* enhanced carrier mobility," *Mater. Sci. Eng. B, vol. 134*, pp. 133-137, Oct. 2006.

[30]   S. Pidin, T. Mori, K. Inoue, S. Fukuta, N. Itoh, E. Mutoh, K. Ohkoshi, R. Nakamura, K. Kobayashi, K. Kawamura, T. Saiki S. Fukuyama, S. Satoh, M. Kase and K. Hashimoto, "A novel strain enhanced CMOS architecture using selectively deposited high tensile and high compressive silicon nitride films," In: *International Electron Devices Meeting Technical Digest*, 2004, pp. 9.2.1-9.2.4.

[31]   S. Ito, H. Namba, K. Yamaguchi, Y. Hirata, K. Ando, S. Koyama, S. Kuroki, N. Ikezawa, T. Suzuki, T. Saitoh and T. Horiuchi, "Mechanical stress effect of etch-stop nitride and its impact on deep submicron transistor design," In: *International Electron Devices Meeting Technical Digest*, 2000, pp. 10.7.1-10.7.4.

[32]   K. Ota, K. Sugihara, H. Sayama, T. Uchida, H. Oda, T. Eimori, H. Morimoto and Y. Inoune, "Novel locally strained channel technique for high performance 55nm CMOS," In: *International Electron Devices Meeting Technical Digest*, 2002, pp. 2.2.1-2.2.4

[33]   C.H. Chen, T.L. Lee, T.H. Hou, C.L. Chen, C.C. Chen, J.W. Hsu, K.L. Cheng, Y.H. Chiu, H.J. Tao, T. Jin, C.H. Diaz, S.C. Chen and M.S. Liang, "Stress memorization technique by selectively strained-nitride capping for sub-65nm high-performance strained-Si device application," In: *Symposium on VLSI Tehnology Digest of Technical Papers*, 2004, pp. 56-57.

[34]   R.E. Belford, B.P. Guo, Q. Xu, S. Sood, A.A. Thrift, A. Teren, A. Acosta, L.A. Bosworth and J.S. Zeil, "Strain enhanced p-type metal oxide semiconductor field effect transistors," *J. Appl. Phys.*, vol. 100, pp. 064903-1-7, Sep. 2006.

[35]   International roadmap for semiconductors, 2010 update. [online] Available: www.itrs.net. [Accessed March 7, 2011]

[36]   T.I. Kamins, *Polycrystalline Silicon for Integrated Circuit Applications,* Kluwer Academic: New York, 1988.

[37]   A.C. Diebold, "The impact of nano-sized dimensions on characterization and metrology," In: *International Conference on Frontiers of Characterization and Metrology for ULSI Technology*, 2007.

CHAPTER 2

Strain, Stress and Semiconductor Properties

Abstract: This tutorial chapter provides a synopsis of the theoretical concepts needed for understanding subsequent chapters. The concepts of engineering strain, true strain, Eulerian strain and in particular, the tensor nature of strain are first introduced. The components of the stress tensor which give rise to strain are next discussed. The three main elastic properties of solids namely: Young's modulus, shear modulus and Poisson's ratio are defined. The piezoresistance effect introduced in the first chapter is discussed quantitatively. This is followed by two topics on the optical properties of semiconductors, the Kramers-Kronig relation and the optical joint density of states used in optical strain metrology.

Keywords: Strain, Stress, Tensor, Piezoresistivity, Dielectric function, Critical points.

1. STRAIN

All solids when being stretched or compressed will undergo deformation. This deformation can be a change in the linear dimensions or a change in the shape of the solid. For most metals and semiconductors, the changes are very small but measurable. On the other hand, for some polymeric materials, the deformation can be very large and is visible to the naked eye. A simple example of this is rubber. The concept of strain is used to describe quantitatively the deformation of solids. It is based on another simple concept called the displacement. Suppose we consider an arbitrary point within a solid bar such as the end point (Fig. **1a**). This point will have a position relative to some frame of reference that is expressed as a set of coordinates. If the bar is subjected to tension along its axis, it will elongate and the point at the end of the bar will be at a new position with a different coordinate. This change in the location of the same point in the solid is called the displacement and strain is basically the normalized displacement of a point within the solid. As a result of this, strain is a dimensionless quantity.

Although the concept of strain is simple to comprehend, its formal expression in mechanics is not straightforward. This is because strain belongs to a class of physical quantities that are called tensors. The simplest physical quantities are those that can be completely specified by one number such as mass or charge. Then there are directed quantities or vectors for which up to three component values are needed for a full specification of that quantity. Examples include the velocity of an object or an electric field. The component values will depend on the system of coordinates chosen. One can generalize this and arrive at physical quantities for which more than three components are needed because each of the components may be associated with more than one of the coordinate axes in space. There are, in fact, a large number of such quantities especially for anisotropic materials and strain is one example.

The simplest type of strain is one dimensional longitudinal strain as illustrated in Fig. **1a**. Due to the applied tension F, the length of the bar is increased from L_0 to L. The change in length in the direction of the force is therefore, $\Delta L = L - L_0$. The engineering strain, ε_e is defined as [1]:

$$\varepsilon_e = \frac{\Delta L}{L_0} = \frac{L}{L_0} - 1 \qquad (2.1)$$

Here, the change in length is divided by the original length. If the change in length is divided instead by the final length, the strain is called the Eulerian strain. For both these definitions, a tensile strain is always positive and all compressive strain is negative. ε_e is called the engineering strain because in mechanics of materials, there is a more rigorous definition of strain called the longitudinal true strain ε_t. This is defined as [1]:

$$\varepsilon_t = \int_{L_0}^{L} \frac{dl}{l} = \ln \frac{L}{L_0} \qquad (2.2)$$

where l is the length variable. By substitution, one can readily show that:

$$\varepsilon_t = \ln(1 + \varepsilon_e) \qquad (2.3)$$

Thus, for the typical small values of strains encountered in engineering solids, the difference between ε_e and ε_t is negligible and this is why the simpler engineering strain is normally used.

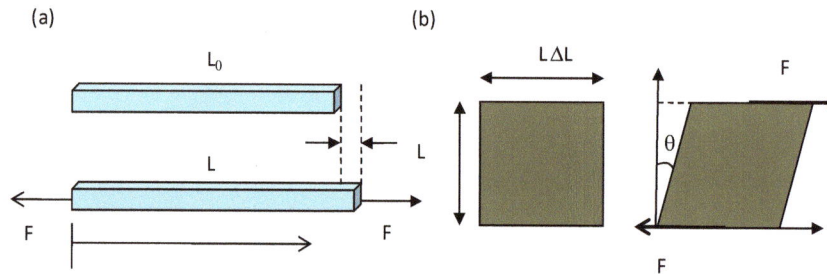

Figure 1: (a) Longitudinal strain and (b) shear strain.

For the above case of longitudinal strain, the shape of the bar remains unchanged. There is another type of strain called the pure shear that involves a change in the shape of the solid but not its linear dimensions. This is illustrated in Fig. **1b**. When a torque (a pair of equal and opposite parallel forces) is applied to the solid on the left, it will be distorted into a parallelepiped. Notice that the applied forces F in this case are acting parallel to the faces of the solid. The shear strain, γ is defined as [1]:

$$\gamma = \frac{\Delta L}{L} = \tan \theta \approx \theta \qquad (2.4)$$

In this definition, ΔL is the displacement in the direction of the force and L is the length perpendicular to ΔL. θ is the angle between ΔL and L. The small angle approximation is often made because the displacement ΔL is usually very small.

In a more general state of strain within a solid, both longitudinal and shear strain can be present and this will be shown in section 4.

2. STRESS

Deformation in a solid occurs because an external force is applied to the solid. For the simplest case of one dimensional longitudinal strain, this force can be applied using a tensile testing machine. After the tensile test specimen is clamped at one end, the other end is pulled by a mechanical force that is increased gradually from zero. For forces that are not sufficient to break the specimen, the specimen will be in static equilibrium and internal forces are induced to keep the specimen in this state. This can be shown by imagining that a perpendicular section is made across some part of the specimen (Fig. **2**). Since the sum of all forces acting on the specimen must be equal to zero, there has to be a force F acting in opposition to the applied force. This internal force is exerted on the other face of the section and is furthermore distributed over the surface of the section. This is because microscopically, it is originated from the chemical bonds between the atoms on both sides of the section. It is thus logical to consider the force per unit cross sectional area which is the normal stress, σ. The normal stress is defined as:

$$\sigma = \frac{F}{A} \qquad (2.5)$$

Here A is the cross sectional area of the specimen when the force F is applied. The normal stress is similar to the true strain in section 1. By analogy, we can also define a simpler engineering stress, σ_e that is based on the original cross sectional area A_0 of the sample (prior to deformation):

$$\sigma_e = \frac{F}{A_0} \tag{2.6}$$

Since the difference between A and A_0 is usually very small, it is generally adequate to use the engineering stress.

For the case of the pure shear, the shear stress, τ is defined similarly as:

$$\tau = \frac{F}{A} \tag{2.7}$$

Here, F is the shear force and A is the area over which the shear force is applied. The direction of the shear force is parallel to the area A. Note that for both normal stress and shear stress, the SI unit is the Pascal (Pa).

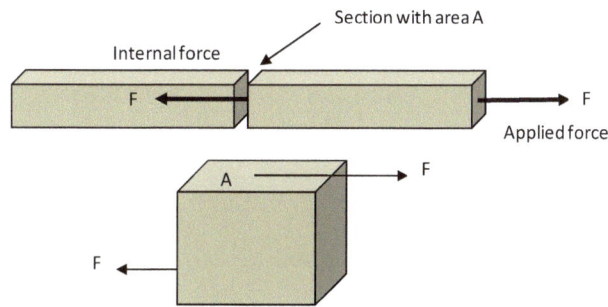

Figure 2: Stress generated within a solid due to applied force.

3. MECHANICAL PROPERTIES OF SEMICONDUCTORS

3.1. Hooke's Law, Young's Modulus and Shear Modulus

The relationship between stress and strain was investigated experimentally by Robert Hooke and in the year 1678, he proposed the law that now bears his name [1]. If we plot the graph of stress versus strain for a bar of semiconductor under tension (Fig. **3**), we will obtain a linear plot in the low stress region. This plot deviates from linearity for higher stresses. The region where the stress-strain curve is linear is also called the elastic regime and is where Hooke's law applies. The elastic regime is characterized by the Young's modulus E which is the ratio of stress and strain or the gradient of the linear stress-strain curve:

$$E = \frac{\sigma}{\varepsilon} \tag{2.8}$$

The Young's modulus is a material property and describes the stiffness of the material. This is because if E is large, then more stress is needed for effecting the same amount of strain. Since strain is dimensionless, E has the unit of stress of the Pascal (Pa) in SI units. Equation (2.8) is also the statement of Hooke's law for one dimensional tensile stress. By analogy, we can write down Hooke's law for the case of the pure shear. This can be written as:

$$G = \frac{\tau}{\gamma} \tag{2.9}$$

where G is the shear modulus or the rigidity. G describes the resistance of the solid to shear distortion and has the same unit as E. The shear modulus is always smaller than the Young's modulus because of the Poisson's ratio to be discussed below. Since both E and G are material properties, they will depend fundamentally on the chemical composition. However, for crystalline materials such as semiconductors, the crystallographic orientation is also very important because the chemical bonding influences both moduli [1].

In the elastic regime, the deformation of the semiconductor is fully reversible. Once the stress is removed, the semiconductor will recover the dimensions and shape that it has before the stress is applied. In this book, we are only concerned with the elastic regime. This is because unlike metals, a semiconductor does not exhibit significant plastic deformation. This refers to the onset of irreversible deformation beyond a characteristic stress called the yield stress. When the applied stress continues to increase, the semiconductor will soon fail by brittle fracture (crack propagation) without showing plastic deformation. The stress at which fracture occurs is called the tensile strength.

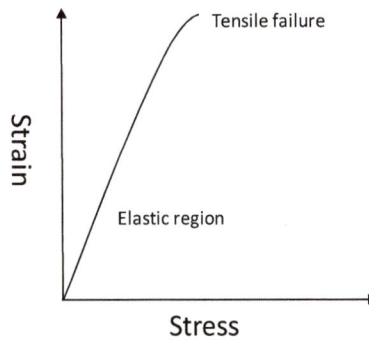

Figure 3: Stress-strain plot for a typical brittle semiconductor such as silicon. Hooke's law is followed until the tensile strength is reached. Fracture occurs at the tensile strength.

3.2. Poisson's Ratio

Fig. **4** shows a more general state of stress at a small cubical volume within a solid. This small volume element is aligned with respect to the three orthogonal axes, x1, x2 and x3. On each of the six faces, there are in general one normal stress and two shear stresses. Due to static equilibrium, there are three independent normal stresses, σ_{11}, σ_{22}, σ_{33} and three independent shear stresses, σ_{12}, σ_{13}, σ_{23}. A double subscript is used to denote each stress component. The first subscript refers to the direction of the normal of the plane on which the force acts. The second subscript refers to the direction of the force itself. Thus, if both subscripts are the same, the stress is acting perpendicular to the plane and is a normal stress. If the subscripts are different, the stress must be acting parallel to a plane and is therefore a shear stress.

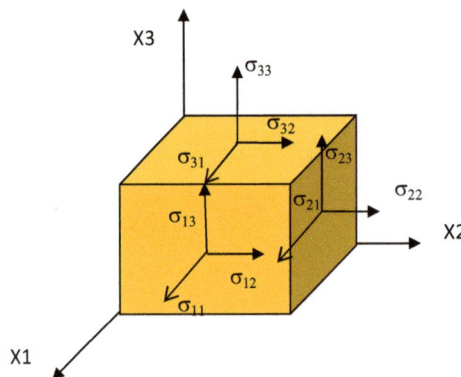

Figure 4: General state of stress showing normal and shear stress components.

The Poisson's ratio is based on the observation that when, say, a normal stress σ_{33} is applied to the solid, it does not only result in a longitudinal strain, ε_{33} but also two transverse (lateral) strains, ε_{11} and ε_{22}. If ε_{33} is tensile, then ε_{11} and ε_{22} are compressive. In other words, elongation in the direction of the third axis is accompanied by contraction in the directions of the other two axes. The Poisson's ratio, υ is dimensionless and is defined as the ratio of the transverse strain to the longitudinal strain when a solid is stressed. For an isotropic solid, the Poisson ratio is independent of direction and we can therefore write:

$$v = -\frac{\varepsilon_{11}}{\varepsilon_{33}} = -\frac{\varepsilon_{22}}{\varepsilon_{33}}$$

(2.10)

There is a minus sign in the definition because we would like the Poisson's ratio to be a positive quantity. Similar relationships can be written for normal stresses applied along the other two axes. By using the relationship between the Young's modulus and the shear modulus, one can prove that the maximum value of υ is 0.5 for an isotropic solid and the minimum value is -1 for the case of a solid with no change in volume [1]. For silicon, the Poisson ratio υ is 0.28.

As a result of Hooke's law and the Poisson's ratio, a normal stress (σ_{11} or σ_{22} or σ_{33}) must give rise to both a longitudinal and a transverse strain. Another way of expressing this is that the strain along any axis ε_{11}, ε_{22}, ε_{33} must take into account all the normal stresses by using the Young's modulus and the Poisson's ratio. For an isotropic solid, this can be written as [1]:

$$\varepsilon_{11} = \frac{1}{E}\left[\sigma_{11} - v\left(\sigma_{22} + \sigma_{33}\right)\right]$$

(2.11)

$$\varepsilon_{22} = \frac{1}{E}\left[\sigma_{22} - v\left(\sigma_{11} + \sigma_{33}\right)\right]$$

(2.12)

$$\varepsilon_{33} = \frac{1}{E}\left[\sigma_{33} - v\left(\sigma_{22} + \sigma_{11}\right)\right]$$

(2.13)

In equations (2.11–2.13), we have made use of the principle of superposition which states that the strain along a direction is the algebraic sum of the contributions due to all three stress components. There is a minus sign within the square bracket because the Poisson's ratio is defined to be a positive quantity. In addition to these three strains, there are three independent shear strains. These are assumed to be due to shear stresses only and so do not involve the normal stresses.

$$\gamma_{12} = \frac{\sigma_{12}}{G}$$

(2.14)

$$\gamma_{13} = \frac{\sigma_{13}}{G}$$

(2.15)

$$\gamma_{23} = \frac{\sigma_{23}}{G}$$

(2.16)

From these relationships, we can see that in the general state of stress of a three dimensional body, there are basically six independent stress components. These are in fact the components of a symmetrical tensor and in the double subscript notation can be written as [1]:

$$\begin{pmatrix} \sigma_{11} & \sigma_{12} & \sigma_{13} \\ \sigma_{12} & \sigma_{22} & \sigma_{23} \\ \sigma_{13} & \sigma_{23} & \sigma_{33} \end{pmatrix}$$

(2.17)

Similarly, there is a symmetrical tensor for the strain and this is in the double subscript notation as [1]:

$$
\begin{pmatrix}
\varepsilon_{11} & \varepsilon_{12} & \varepsilon_{13} \\
\varepsilon_{12} & \varepsilon_{22} & \varepsilon_{23} \\
\varepsilon_{13} & \varepsilon_{23} & \varepsilon_{33}
\end{pmatrix}
\tag{2.18}
$$

Note that the off diagonal terms in this tensor are defined as: $\varepsilon_{12} = \gamma_{12}/2$, $\varepsilon_{13} = \gamma3/2$; $\varepsilon_{23} = \gamma_{23}/2$. This definition often causes confusion to the beginner in tensors. The reason should be apparent after reading the next page. However, we can point out here that the two symmetrical tensors is linked by a higher order tensor with 81 (9x9) elements. If we were to invoke symmetry and reduce this to 36 elements (6x6), then in order to preserve the equations, one must define the off diagonal components of the strain tensor as above. If both tensors are symmetric, there is actually no need to use double subscripts and one can use the following notation instead:

$$
11 \rightarrow 1 \; ; \; 22 \rightarrow 2 \; ; \; 33 \rightarrow 3; \; 23 \rightarrow 4; \; 13 \rightarrow 5; \; 12 \rightarrow 6
\tag{2.19}
$$

Using this single subscript notation, the stress tensor and strain tensor become respectively:

$$
\begin{pmatrix}
\sigma_1 & \sigma_6 & \sigma_5 \\
\sigma_6 & \sigma_2 & \sigma_4 \\
\sigma_5 & \sigma_4 & \sigma_3
\end{pmatrix}
\tag{2.20}
$$

$$
\begin{pmatrix}
\varepsilon_1 & \varepsilon_6/2 & \varepsilon_5/2 \\
\varepsilon_6/2 & \varepsilon_2 & \varepsilon_4/2 \\
\varepsilon_5/2 & \varepsilon_4/2 & \varepsilon_3
\end{pmatrix}
\tag{2.21}
$$

There is a factor of ½ because ε_4, ε_5 and ε_6 are the shear strains. Using this notation, we can finally write down the generalized stress strain relationship as:

$$
\begin{bmatrix}
\sigma_1 \\
\sigma_2 \\
\sigma_3 \\
\sigma_4 \\
\sigma_5 \\
\sigma_6
\end{bmatrix}
=
\begin{bmatrix}
C_{11} & C_{12} & C_{13} & C_{14} & C_{15} & C_{16} \\
C_{21} & C_{22} & C_{23} & C_{24} & C_{25} & C_{26} \\
C_{31} & C_{32} & C_{33} & C_{34} & C_{35} & C_{36} \\
C_{41} & C_{42} & C_{43} & C_{44} & C_{45} & C_{46} \\
C_{51} & C_{52} & C_{53} & C_{54} & C_{55} & C_{56} \\
C_{61} & C_{62} & C_{63} & C_{64} & C_{65} & C_{66}
\end{bmatrix}
\begin{bmatrix}
\varepsilon_1 \\
\varepsilon_2 \\
\varepsilon_3 \\
\varepsilon_4 \\
\varepsilon_5 \\
\varepsilon_6
\end{bmatrix}
\tag{2.22}
$$

Since the above is cumbersome, the shorthand tensor notation is used instead:

$$
\sigma_i = C_{ij}\varepsilon_j
\tag{2.23}
$$

where C_{ij} is called the stiffness ($k, j = 1...6$). The stress strain equation can be written alternatively as:

$$
\begin{bmatrix}
\varepsilon_1 \\
\varepsilon_2 \\
\varepsilon_3 \\
\varepsilon_4 \\
\varepsilon_5 \\
\varepsilon_6
\end{bmatrix}
=
\begin{bmatrix}
S_{11} & S_{12} & S_{13} & S_{14} & S_{15} & S_{16} \\
S_{21} & S_{22} & S_{23} & S_{24} & S_{25} & S_{26} \\
S_{31} & S_{32} & S_{33} & S_{34} & S_{35} & S_{36} \\
S_{41} & S_{42} & S_{43} & S_{44} & S_{45} & S_{46} \\
S_{51} & S_{52} & S_{53} & S_{54} & S_{55} & S_{56} \\
S_{61} & S_{62} & S_{63} & S_{64} & S_{65} & S_{66}
\end{bmatrix}
\begin{bmatrix}
\sigma_1 \\
\sigma_2 \\
\sigma_3 \\
\sigma_4 \\
\sigma_5 \\
\sigma_6
\end{bmatrix}
\tag{2.24}
$$

Again in shorthand tensor notation,

$$
\varepsilon_i = S_{ij}\sigma_j
\tag{2.25}
$$

where S_{ij} is called the compliance. Both S_{ij} and C_{ij} are fourth order tensors. Mathematically, there should be 36 components for a fourth order tensor. However, for a semiconductor with a cubic unit cell such as silicon, the symmetry of the unit cell will reduce the number of non-zero tensor components to just three. These are: C_{11} = 167 GPa; C_{12} = 65 GPa and C_{44} = 80 GPa [2]. The values for other cubic semiconductors of technological importance can be found appendix A2.

4. PIEZORESISTIVITY

The peizoresistivity effect was discovered by Lord Kelvin in the 19th century. In a conductor which does not exhibit piezoresistivity, the electric field is simply linearly related to the current density *via* the resistivity by Ohm's law. In a piezoresistive material, the electric field depends on the external stress, X_{kl} as well as the current density, I_j. This can be expressed as [3]:

$$E_i = f(I_j, X_{kl}) \qquad (2.26)$$

The four indices i,j,k,l have the values 1,2 and 3 representing three orthogonal directions. In order to obtain the functional form f, we can use a McClaurin's series expansion of (2.27) about zero stress and zero current. This is because we are only interested in very small deformations. The differential change in the electric field then becomes [3]:

$$dE_i = \left(\frac{\partial E_i}{\partial I_j}\right)dI_j + \left(\frac{\partial E_i}{\partial X_{kl}}\right)dX_{kl} + \frac{1}{2!}\left[\left(\frac{\partial^2 E_i}{\partial I_j \partial I_m}\right)dI_j I_m + \left(\frac{\partial^2 E_i}{\partial X_{kl}\partial X_{no}}\right)dX_{kl}dX_{no} + 2\left(\frac{\partial^2 E_i}{\partial X_{kl}\partial I_j}\right)\right] + ... \quad (2.27)$$

The first partial derivative on the right hand side of (2.27) is the familiar resistivity tensor ρ_{ij}. The second partial derivative $\left(\partial E_i / \partial X_{kl}\right)$ is called the converse piezoelectric tensor d_{ikl} and the third partial derivative $\left(\partial^2 E_i / \partial I_i \partial I_m\right)$ is the nonlinear resistivity tensor ρ_{ijm}. The fourth partial derivative is the nonlinear piezoelectric tensor δ_{ijlno} and the last partial derivative $\left(\partial^2 E_i / \partial X_{kl}\partial I_j\right) = \left(\partial / \partial X_{kl}\right)\left(\partial E_i / \partial I_i\right) = \Pi_{ijkl}$ is the piezoresistivity tensor. Using these tensors and after integration, we can write the electric field as:

$$E_i = \rho_{ij}I_j + d_{ikl}X_{kl} + \frac{1}{2}\left(\rho_{ijm}I_j I_m + \delta_{ik\ln o}X_{kl}X_{no}\right) + \Pi_{ijkl}X_{kl}I_j \qquad (2.28)$$

As a result of centrosymmetry in most technologically important semiconductors including silicon and germanium [3], the only term that will contribute to change in the electric field under stress is the piezoresistive term. Thus equation (2.28) can be further simplified for these semiconductors.

As with the stiffness and compliance tensors in Hookes' law, the piezoresistivity tensor is really a tensor of rank 4 and thus should have 81 tensor components. However, for the triclinic crystal structure, symmetry reduces the number of independent tensor components to only 36 [3]. Thus, once again the matrix notation described in section 2.3 can be used to simplify the piezoresistivity tensor. For the cubic semiconductors such as silicon and germanium, further symmetry considerations reduce the number of independent components to just three and these are Π_{11} Π_{12} Π_{44} with all other components of the matrix equal to zero as shown below:

$$\Pi_{mn} = \begin{bmatrix} \Pi_{11} & \Pi_{12} & \Pi_{12} & 0 & 0 & 0 \\ \Pi_{12} & \Pi_{11} & \Pi_{12} & 0 & 0 & 0 \\ \Pi_{12} & \Pi_{12} & \Pi_{11} & 0 & 0 & 0 \\ 0 & 0 & 0 & \Pi_{44} & 0 & 0 \\ 0 & 0 & 0 & 0 & \Pi_{44} & 0 \\ 0 & 0 & 0 & 0 & 0 & \Pi_{44} \end{bmatrix} \qquad (2.29)$$

5. OPTICAL INTERACTIONS OF SEMICONDUCTORS

In this section, we discuss in a concise manner the optical properties of semiconductors. A basic knowledge of semiconductor optical properties and their relationship with the electronic band structure of a semiconductor is essential to understanding the optical strain metrology techniques to be described in part two. We shall introduce a number of semiconductor optical properties that are normally only found in more advanced texts on semiconductors. However, since this is not a textbook on semiconductors, the coverage will be necessarily selective. Readers who would like a broader exposition of this important field should refer instead to a number of excellent texts on this topic [4, 5].

When electromagnetic radiation such as visible light is incident on a semiconductor, the photons in the incident light can interact with the semiconductor in a number of ways. This is schematically illustrated in Fig. **5**. The first possible interaction is specular reflection which is similar to what happens when light is incident on a polished metallic surface. The incident photons do not enter the semiconductor and are instead returned to the ambient. The angles of incidence and reflection are equal. This is due to the large refractive index difference of the semiconductor and ambient. By applying Maxwell's electromagnetic theory, one can determine the fraction of the incident photons or incident energy that is reflected using the optical properties of the semiconductor.

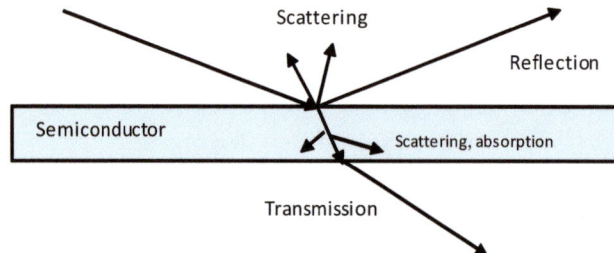

Figure 5: Possible interactions of light with a semiconductor substrate.

The remaining fraction of the incident light is transmitted into the semiconductor. Once inside the semiconductor, the transmitted light can interact with the semiconductor in several ways as shown in Fig. **5**. First, the light can go through the semiconductor completely without being absorbed. This occurs whenever the semiconductor is transparent to the incident light and this means the energy of the photons is less than the fundamental energy bandgap of the semiconductor. This for silicon is about 1.1eV and corresponds to the near infra-red region of the electromagnetic spectrum. Although transmission can readily occur in semiconductor thin films, the transmitted photons can also be scattered by the semiconductor crystal prior to leaving the crystal. When a photon is scattered, the direction of propagation of the photon and sometimes its energy is changed by the scattering interaction. Several types of scattering can occur namely: (i) Rayleigh scattering; (ii) Raman scattering and (iii) Brillouin scattering. Rayleigh scattering is elastic scattering and only the direction of the photon or momentum is changed as a result. It occurs because of lattice interactions and is a dominant form of light scattering in semiconductors.

Both Raman and Brillouin scattering are inelastic scattering events. For these two types of scattering, the incident photon can either lose or gain some characteristic amount of energy from the semiconductor. Hence both the energy and direction of the scattered photon are different from the incident photon unlike the Rayleigh case. Raman scattering forms the basis of the micro-Raman spectroscopy technique to be discussed in chapter 5. In this case, the energy change of the scattered photon is due to the interaction with the optical phonons or high frequency lattice vibrations of the semiconductor. The energy of the phonon modes is quantized like the energy of the electronic states and is specific to each semiconductor. Hence, by measuring the change in the photon energy or frequency (Raman shift), one can characterize the semiconductor. Brillouin scattering is similar to Raman scattering but the scattering phonons are acoustic or low frequency phonons. At present, Brillouin scattering is not used for monitoring strain in semiconductors and so will not be discussed further in this eBook.

The last and important optical interaction is absorption. In this process, the energy of the incident photon is used to break chemical bonds in a semiconductor. An electron is excited from the valence band to the conduction band to form an electron hole pair. Absorption can occur whenever the energy of the incident photon is greater than the energy bandgap of the semiconductor. However, it is more likely to occur in a direct bandgap semiconductor where the bottom of the conduction band and the top of the valence band has the same wavevector, k. Absorption is the fundamental mechanism behind all semiconductor photodetectors, charge coupled devices in digital cameras and the photovoltaic device. In strain metrology, absorption is also important because the energy at which direct interband absorption occurs is dependent on the lattice strain. Thus, by monitoring the photon energy at which this absorption occurs will enable one to deduce the strain of the semiconductor. This is discussed in chapter 6.

6. OPTICAL PROPERTIES OF SEMICONDUCTORS

The optical interactions discussed above can be predicted if the optical properties of the semiconductor are known. The optical properties in general are tensor quantities because they can depend on the orientation within the crystal. For clarity, it is assumed in the following that the medium is isotropic and so the optical properties can be written like scalars. For all semiconductors, the optical properties can be expressed in two equivalent ways as function of energy E as: (i) the complex dielectric function, $\varepsilon(E)$ and (ii) the complex refractive index, $n*(E)$. Both the complex dielectric function and the complex refractive index have a real part and an imaginary part and following the customary notation in optics, they will be written as [5]:

$$\varepsilon\left(E\right)=\varepsilon_{1}\left(E\right)+i\varepsilon_{2}\left(E\right) \tag{2.30}$$

$$n*\left(E\right)=n\left(E\right)+ik\left(E\right) \tag{2.31}$$

where $\varepsilon_{1}=\mathrm{Re}(\varepsilon)$ and $\varepsilon_{2}=\mathrm{Im}(\varepsilon)$ are the real and imaginary parts of $\varepsilon(E)$ and n and k are the refractive index and extinction coefficient of n respectively. The scalar, $i=\sqrt{-1}$. A complex function notation is used to represent the optical properties because the incident light can both propagate through the semiconductor and be absorbed by the semiconductor. For a sinusoidal incident wave, it is convenient to use the phasor notation to describe these two effects and in this way, the optical properties can be mathematically represented by the real and imaginary parts of a complex dielectric function or complex refractive index.

There are two important sets of relationships for the optical properties of a semiconductor. The first concerns the relationship between ε(E) and n*(E).

$$n*\left(E\right)=\sqrt{\varepsilon\left(E\right)} \tag{2.32}$$

By substituting from (2.30) and (2.31), one readily finds that:

$$\varepsilon_{1}=n^{2}-k^{2} \tag{2.33}$$

$$\varepsilon_{2}=2nk \tag{2.34}$$

Thus, if n and k are known or measured, ε_{1} and ε_{2} can be calculated immediately. In addition, by considering n and k to be unknowns and solving for them quadratically, one can express n and k using the real and imaginary parts of the complex dielectric function.

The second important relationship is called the Kramers-Kronig (K-K) relationship. The K-K relationship is needed to understand both the spectroscopic ellipsometric method (chapter 3) and the photoreflectance method (chapter 4) for strain metrology. This states that the real and imaginary parts of $\varepsilon(E)$ or $n*(E)$ are in fact inter-related. For the complex dielectric function, it can be proved that [4]:

$$\varepsilon_1(E) = 1 + \frac{2}{\pi} \int_0^\infty \frac{E'\varepsilon_2(E)}{E'^2 - E^2} dE' \tag{2.35}$$

$$\varepsilon_2(E) = -\frac{2E}{\pi} \int_0^\infty \frac{E'\varepsilon_1(E)}{E'^2 - E^2} dE' \tag{2.36}$$

In the above relations, E' is a dummy integration variable with the unit of energy. If the value of ε_2 is known at the energy E, then integration over the entire energy range with respect to the dummy variable E' needs to be performed to find the value of ε_1 at energy E. The same type of integration is needed to find ε_2 from ε_1. Similarly, for the complex refractive index, it can shown that [5]:

$$n(E) = 1 + \frac{2}{\pi} \int_0^\infty \frac{E'k(E)}{E'^2 - E^2} dE' \tag{2.37}$$

$$k(E) = -\frac{2E}{\pi} \int_0^\infty \frac{E'n(E)}{E'^2 - E^2} dE' \tag{2.38}$$

This means that once the real part of the complex dielectric function or complex refractive index is known, the imaginary part can be calculated and vice versa.

It is important to emphasise that both $\varepsilon(E)$ and $n*(E)$ are functions of the photon energy or frequency and this can span a wide spectral range. The optical functions of a semiconductor can be measured from the very low frequency (Restrahlen) region through the infrared and visible region to the extreme ultra-violet regions. For strain metrology, it is the optical properties near the fundamental energy gap that are really important. For silicon, this occurs in the visible light region. In the following, we will therefore focus on the optical properties near the absorption edge.

7. JOINT DENSITY OF STATES AND CRITICAL POINTS

When the energy of incident light is equal to or greater than the optical bandgap of a semiconductor, photon absorption will occur. In the optical functions, this is manifested as increased values of ε_2 and k. Both these components of the respective optical functions have negligible values when the photon energy is far below the optical bandgap. It is important to understand the origin of $\varepsilon_2(E)$ at energies near the optical bandgap. Although $\varepsilon_2(E)$ can be found by the K-K relations, fundamentally, $\varepsilon_2(E)$ is determined by the electronic band structure of the semiconductor as shown by the following equation [6]:

$$\varepsilon_2(E) = \frac{4e^2\hbar^2}{\pi\mu^2 E^2} \int |P_{cv}(k)|^2 \delta(E_c(k) - E_v(k) - E) dk \tag{2.39}$$

In this equation, $E_c(k)$ and $E_v(k)$ are the band structure (E - k relation) of the conduction and valence bands respectively. The Dirac delta function, $\delta(E)$ is called the spectral joint density of states (JDOS) of the conduction and valence bands separated by the energy E where E is the photon energy. The JDOS function yields the density of available states at the conduction and valence bands for absorption when it is multiplied by the matrix element $P_{cv}(k)$ of the conduction and valence bands and integrated over the first Brillouin zone in k space.

The integral above is a volume integral because $dk = dk_x dk_y dk_z$. It can be mathematically converted into a surface integral with respect to the surface element ds by using vector calculus:

$$\varepsilon_2(E) = \frac{4e^2\hbar^2}{\pi\mu^2 E^2} \int \frac{|P_{cv}(k)|^2 dS}{|\nabla_k(E_c - E_v)|\big|_{Ec-Ev=E}} \tag{2.40}$$

For most semiconductors, $P_{cv}(k)$ is not a strongly varying function [6]. Thus, it can be taken outside the surface integral and what remains is defined as the JDOS function:

$$J_{cv}(E) = \int \frac{dS}{\left|\nabla_k (E_c - E_v)\right|_{Ec-Ev=E}} \tag{2.41}$$

This definition has the advantage that its meaning is clearer than the Dirac function mentioned above. The JDOS function is the surface integral of the reciprocal of the gradient function of the difference between the conduction and valence bands. This surface integral has to be evaluated over a constant energy surface for which the difference between E_c and E_v is equal to E. In other words, from the energy band structure of the semiconductor, one first obtains the energy difference $E_c - E_v$ and construct the constant energy surface with energy E in k space. Then the gradient function of the scalar, $E_c - E_v$ needs to be found and this is used to obtain $J_{cv}(E)$. Since the gradient function is given by the derivative of $E_c - E_v$ with respect to k_x, k_y and k_z, it is immediately obvious that the joint density of states function should have points where the gradient function is zero and the integral becomes infinite. These special points or singularities in the JDOS function all satisfy the condition:

$$\nabla_k \left| E_c(k) - E_v(k) \right| = 0 \tag{2.42}$$

This can also be written as:

$$\nabla_k E_c(k) = \nabla_k E_v(k) = 0 \tag{2.43}$$

These special points in the JDOS function are called the critical points (Fig. **6**). This figure summarizes the four types of critical points and the shape of the JDOS function near these points.As will be seen in chapters 3 and 4, the critical points are the cornerstone of optical strain metrology in semiconductors. By considering the sign of the second derivative of the function $E_c - E_v$, it can shown further that there are actually four types of critical points and these are called the minimum point M_0, maximum point M_3 and the saddle points, M_1 and M_2 [6].

Note that the critical points of the JDOS function are associated with turning points and saddle points in the band structure of the semiconductor. This can be seen by referring to appendix 1a which is the computed *E-k* diagram for unstrained silicon along several directions defined by points of high symmetry in the first Brillouin zone. Four critical point transitions are annotated on this diagram. Each one is a direct transition and they are all greater in energy than the fundamental gap or indirect transition in silicon. At the Γ point, there is one transition called $E_0{'}$ in the literature. In addition, there are two transitions at the L point are called E_1 and $E_1{'}$ respectively. The transition at the X point is called E_2. These transitions all correspond to critical points in the JDOS function. The E_1 critical point, in particular, is used in the spectroscopic ellipsometry method for determining strain in strained silicon in chapter 3.

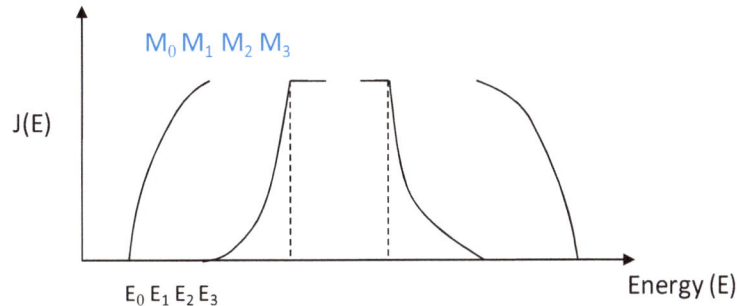

Figure 6: Joint density of states function and the four major types of critical points.

REFERENCES

[1]   M.A. Meyers and K.K. Chawla, *Mechanical Properties of Materials,* Cambridge University Press: Cambridge, 2009.

[2]   S. Adachi, *Properties of Group-IV, III-V and II-VI Semiconductor,* Wiley: Chichester, 2005.

[3]   R.C. Dorf, *The Electrical Engineering Handbook,* CRC Press: Boca Raton, 1995.

[4]   N.W. Ashcroft and N.D. Mermin, *Solid State Physics*, Saunders College: Philadelphia, 1976.

[5]   C. Kittel, *Introduction to Solid State Physics*, seventh edition, Wiley: New York, 1996.

[6]   J.I Pankove, *Optical Processes in Semiconductors,* Dover: New York, 1971.

PART 2: OPTICAL STRAIN METROLOGY

In this part, we describe three strain metrology techniques that make use of the interaction of light with a sample to measure strain. The optical interactions are: (i) reflection, (ii) polarization change and (iii) inelastic scattering. Although the interactions used are different for each technique, they share a common principle that they are based on the so-called morphic effects of semiconductors. The morphic effects are basically perturbation effects by external influences such as stress or pressure on the crystal structure and the band structure of a semiconductor. By measuring the changes in the band structure, it is possible for instance to infer the magnitude of the strain that causes the perturbation.

All three techniques make use of visible and ultraviolet light and are non-destructive and non-invasive. They therefore involve little or no sample preparation and are well suited to the routine process monitoring of strained semiconductor substrates such as strained silicon and strained silicon on insulator. However, the techniques are mainly *ex situ* and with the exception of the Raman spectroscopy technique cannot be applied to individual transistors on wafer. The tip-enhanced Raman technique with higher spatial resolution is at present under development and is not yet ready for manufacturing.

CHAPTER 3

Variable Angle Spectroscopic Ellipsometry

Abstract: Strain measurement based on reflection of polarized light from the strained semiconductor sample is explained in this chapter. Spectroscopic ellipsometry is usually applied for thickness measurement of thin films of materials with known optical functions. In this approach, the optical functions are found with the layer thickness determined from additional separate measurements. From the fitted optical function spectra, the shift in energy of the critical point can be determined mathematically and using the deformation potential, the strain can be obtained.

Keywords: Spectroscopic ellipsometry, Piezo-optical effect, Parametric semiconductor model, Deformation potential, Virtual substrate.

1. INTRODUCTION

The method of variable angle spectroscopic ellipsomery (SE) was proposed by the author in 2007 as a possible solution to the in-line measurement problem of strained silicon wafers [1]. SE is a widely used optical characterization technique that is used in every wafer fabrication facility for process monitoring. Vineis *et al.* [2] had shown that SE can accurately and reproducibly determine the thickness of ε-Si layers grown on SiGe virtual substrates (VS). SE was also used by Schmidt *et al.* to determine the thickness of ε-Si and the Ge composition of the SiGe VS [3]. Fursenko *et al.* had also used SE as an in-line process control tool for SiGe:C heterojunction bipolar transistors [4].

In ε-Si, elastic strain due to lattice-mismatch between Si and SiGe causes a tetragonal deformation of the unstrained cubic Si lattice. The reduced crystal symmetry results in a partial lifting of the degeneracies at the bottom of the conduction band and the top of the valence band [5]. In addition, both the electron and hole effective mass are altered by changes in the curvature of the Si band structure [6]. This together with the reduced inter-valley scattering results in enhanced carrier mobilities. For example, in ε-Si with 1% tensile strain, an 80% increase in electron field effect mobility at 0.1 MV/cm has been observed [5]. Hole mobility enhancement has also been observed in tensile ε-Si. However, it was found to degrade at high gate voltage overdrive [5]. Since mobility enhancement in ε-Si is due solely to the presence of strain, strain monitoring during fabrication and strain relaxation are both of great practical concern.

Due to the above reasons, it is desirable to have an in-line characterization approach capable of monitoring both the strain and film thickness during ε-Si device manufacturing. Alternative techniques for strain characterization include [7]: transmission electron microscopy/diffraction (TEM), high resolution X-ray diffraction (HR-XRD), Raman spectroscopy and ion-channeling [8]. (These will be discussed in subsequent chapters). TEM based electron diffraction is destructive and can only be applied to small samples. During sample preparation, the strain field in the ε-Si can be modified. Although in-line HR-XRD instruments are now available, their measurement time is generally long. Ion channeling requires the use of high energy ion accelerators at specialized laboratories. Thus an alternative non-destructive strain characterization technique is needed.

In this chapter, we demonstrate that SE together with X-ray scattering can be used to determine strain in ε-Si. The approach is based on the determination of the complex dielectric functions of ε-Si and its representation in a compact parametric form. The shifts in the critical point (CP) energies of the dielectric function spectra are then related to the strain through known mathematical relationships. It will be seen that this method is in some ways complementary to the photoreflectance method that will be introduced in chapter 4. However, unlike this method, use of the SE approach will not involve purchase of new equipment for a wafer fabrication facility.

2. SE STRAIN MEASUREMENT PRINCIPLE

The SE method is basically an optical measurement utilizing polarized light [9]. The broadband light source used in SE is typically in the visible, near infrared and ultraviolet (UV) spectral region. However, in recent years, deep UV light is also being used. The light can be either reflected or transmitted by the sample. Since strained silicon is not transparent to visible light, a reflection ellipsometer is used in this chapter. After passing through a polarizer in the ellipsometer, the incident light can become linearly polarized. This means the oscillating plane of the electric field has a fixed orientation with respect to the plane of reflection. The latter is the plane containing the incident beam, reflected beam and surface normal and is therefore always perpendicular to the sample. The electric field vector E_i of linearly polarized light can be resolved into two components that are parallel and perpendicular to the incident plane. The parallel component is called E_{pi} and the perpendicular component E_{si}. Both are complex (phasor) quantities with a magnitude and phase. After reflection from the sample, the polarized light can also be resolved into two components, E_{pr} and E_{sr} but the amplitude and phase of these complex components will be different from the incident light. If we divide the magnitude of corresponding components of the reflected and incident light, we obtain the magnitude of the reflection coefficient.

$$|R_p| = \frac{|E_{pr}|}{|E_{pi}|} \tag{3.1}$$

$$|R_s| = \frac{|E_{sr}|}{|E_{si}|} \tag{3.2}$$

The ratio of these two magnitudes of the reflection coefficients R_p and R_s is by definition called the tan ψ, where ψ is an angle.

$$\tan \psi = \frac{|R_p|}{|R_s|} \tag{3.3}$$

The two components of the incident wave E_{pi} and E_{si} has a phase difference between them that can be called δ_1. Similarly, the reflected wave components have a phase difference that is δ_2. The difference between these two phase differences is called the delta Δ:

$$\Delta = \delta_1 - \delta_2 \tag{3.4}$$

All ellipsometric measurements is about solving the following ellipsometric equation [9]:

$$\rho = \frac{R_p}{R_s} = \tan \psi \bullet \exp(i\Delta) \tag{3.5}$$

In this equation, ρ is the complex ratio of the two complex reflection coefficients R_p and R_s and should not to be confused with the resistivity. As shown by equation (3.5), the two angles ψ and Δ will specify this complex ratio. If this is repeated for each wavelength of the incident light, then one will obtain the spectra of ψ and Δ. A spectroscopic ellipsometer is basically an instrument used for measuring this type of spectra. After measurement, a parametric model of the sample needs to be developed. By performing mathematical fitting of the ψ and Δ spectra with this parametric model of the sample, one can obtain sample information such as the thickness of the layers when a good fit between the measured and simulated ψ and Δ spectra are obtained. This in fact had been the usual application of the SE technique. In this conventional application, the optical functions of the material are usually known and are used as input. However, for strain metrology, we approach this in the opposite way. We measure the sample thickness by other independent methods and use these as input to find the unknown optical functions of the strained sample. Once the optical functions are obtained by fitting, we can analyze these functions to obtain the critical points. This is discussed next.

In principle, the strain of ε-Si can be found optically *via* the piezo-optical effect which is given by equation (3.6) [10]:

$$\Delta\varepsilon_{ij}(\omega) = P_{ijkl}(\omega) X_{kl} \tag{3.6}$$

Here, $\Delta\varepsilon_{i\varphi}(\omega)$ is the change of the dielectric function tensor, $P_{ijkl}(\omega)$ is the piezo-optical tensor and X_{kl} is the applied stress tensor. $\varepsilon_{i\varphi}(\omega)$ and X_{kl} are both tensors of rank 2 while $P_{ijkl}(\omega)$ is of rank 4. Since silicon has cubic symmetry, there are only three independent components of $P_{ijkl}(\omega)$, namely $P_{11}(\omega)$, $P_{12}(\omega)$ and $P_{44}(\omega)$. The values of these have been measured previously [10]. Thus, by measuring the components of the change in the dielectric function tensor, the stress in the ε-Si can be deduced. If Si is deformed elastically, Hooke's law can be applied to determine the strain tensor by using the tabulated elastic constants of Si.

In practice, however, measurement of the piezo-optical tensor requires a spectroscopic ellipsometer equipped with a stage for applying stress to the sample. Several ellipsometric spectra had to be collected in directions parallel and perpendicular to the applied stress direction and the sample is restricted to bulk single crystal form of the material being investigated [10]. These requirements mean that the method may not be suitable for semiconductor thin film characterization.

In this work, we adopted an indirect approach to optically monitor the strain in ε-Si. The complex dielectric function of crystalline Si over the energy range 2.0-5.5eV is characterized by several interband critical points (CP) designated as E_0', E_1, E_2 and E_1' [11]. E_0' originates from direct transitions from the top of the valence band (VB) to the bottom of the conduction band (CB) at the Γ point of the Brillouin zone (BZ) while E_1 is due to transitions from the VB to the CB in the Λ direction near the L point (see appendix 1). At room temperature, these two CPs are nearly degenerate at ~3.4eV with an energy separation of ~70meV [11]. The critical point E_2 is located at higher energy and has been assigned to transitions near the X point in the Δ direction. The next CP E_1' involves direct excitation from the VB to the second lowest CB at the L point of the BZ.

As observed experimentally by Lee and Jones [12], the effect of biaxial stress on the dielectric function of Si consists of a strain induced splitting of the E_1 CP into E_1 and $E_1+\Delta_1$. In addition, hydrostatic strain decreases (increases) the energy of these points for tensile (compressive) stress. This results in an apparent down shift of the E_1 edge. The strain effects on E_1 is described by the following equation [13]:

$$E_\pm = E_1(0) + \delta_H + \frac{\Delta_1}{2} \pm \frac{1}{2}\sqrt{\Delta_1^2 + 4\delta_s^2} \tag{3.7}$$

where the plus sign refers to $E_1+\Delta_1$ and minus sign refers to E_1. $E_1(0)$ is the energy of E_1 in the absence of strain and Δ_1 is the spin-orbit splitting. δ_H is the change in CP energy due to hydrostatic strain and δ_s is the change due to uniaxial shear strain. δ_H and δ_s are given respectively by [13]:

$$\delta_H = \frac{2D_1^1}{\sqrt{3}}\left(1 - \frac{C_{12}}{C_{11}}\right)\varepsilon_{xx} \tag{3.8}$$

$$\delta_S = -\sqrt{\frac{2}{3}}D_3^3\left(1 + \frac{2C_{12}}{C_{11}}\right)\varepsilon_{xx} \tag{3.9}$$

Here, D_1^1 and D_3^3 are the hydrostatic and shear deformation potential respectively. The deformation potentials describe the change in energy of an electronic state per unit strain. C_{11} and C_{12} are the elastic constants of Si and ε_{xx} is the in-plane strain. By using published values of the deformation potentials [14] and elastic constants of Si in literature [15], a graph of the two strain dependent terms in (2) *versus* strain can be plotted (Fig. **1**). As shown, δ_H is a linear function of ε_{xx} while the other term involving δ_S tends to a constant value at small values of ε_{xx}. Using this data, it can be shown that the $E_1 + \Delta_1$ CP is shifted up slightly in energy (~22meV) with respect to $E_1(0)$ and it is not sensitive to strain. On the other hand, E_1 is shifted to lower energy and the shift increases with strain. Although there are variations in the reported values for the deformation potentials [13], the above conclusion is not fundamentally affected by this. If the position of this peak in the dielectric function spectrum is found by SE, the strain ε_{xx} of the ε-Si can be determined.

As pointed out by Zollner *et al.* [16], the method is complicated by the proximity of the $E_0{}'$ and E_1 CPs at room temperature. Since the separation is close to the peak broadening, it is difficult to separate these two peaks [16]. Low temperature SE may be needed to resolve these two CPs. Furthermore, the effect of biaxial stress on $E_0{}'$ has not been reported in literature. An implicit approximation in the measurement method described above is the use of an isotropic dielectric function for ε-Si. The dielectric function of ε-Si is strictly anisoptropic with an ordinary (in-plane) and extraordinary (out of plane) component. However, as pointed out by Vineis *et al.* [2], this birefringence of ε-Si can be ignored because the measured SE data is dominated by the ordinary component.

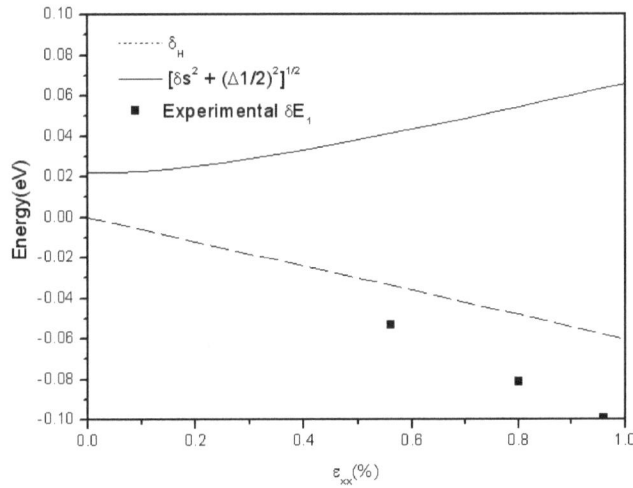

Figure 1: Hydrostatic and shear components of the energy of the E_1 and $E_1 + \Delta_1$ critical points as function of strain. Observed experimental shifts and sample strains are included [1]. Reprinted with permission 'Characterization of biaxial stressed silicon by spectroscopic ellipsometry and synchrotron X-ray scattering', T.K.S. Wong, Y. Gong, P. Yang, C.M. Ng, Semicond. Sci, and Technol. 22, 1232-1239, November 2007, IOP Publishing. DOI: 10.1088/0268-1242/22/11/009.

3. EXPERIMENTAL METHOD

For the demonstration of this method, three ε-Si samples (S15, S20, S25) were grown epitaxially on $Si_{1-x}Ge_x$ VS by chemical vapor deposition. The VS grown on (100) Si wafers consists of a 2μm graded $Si_{1-x}Ge_x$ buffer layer with final Ge compositions of 15 – 25 at.% and a 1μm thick uniform $Si_{1-x}Ge_x$ cap layer with the same final Ge composition as the top of the graded $Si_{1-x}Ge_x$ buffer layer. The nominal sample characteristics are summarized in Table **1**.

Table 1: Characteristics of Biaxial ε-Si Epitaxial Samples

	S15	S20	S25
ε-Si thickness (nm)	20	20	25
Si_xGe_{1-x} cap layer thickness (μm)	1	1	1
Final Ge concentration x of buffer layer (at.%)	15	20	25
Si_xGe_{1-x} buffer layer thickness (μm)	2	2	2

SE data from 250nm – 500nm was collected using a J.A. Woollam variable angle spectroscopic ellipsometer (VASE) with a rotating analyzer and automatic goniometer (Fig. **2**). The angle of incidence (AOI) was varied from 75° to 81°. At the optimized AOI for these samples of ~78° [9], the SE sensitivity to the ε-Si samples is maximized. All SE measurements were conducted in air at room temperature. Data fitting was performed using the software WVASE32.

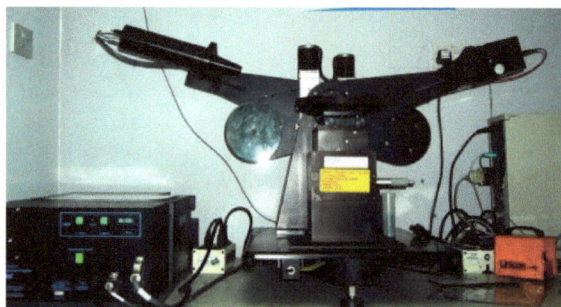

Figure 2: Photograph of a laboratory variable angle spectroscopic ellipsometer for strained silicon measurements.

In order to facilitate fitting of the SE data, additional measurements were carried out using synchrotron X-rays. More information about the beamline can be found in chapter 11. The energy of the photons used in this work was 8keV (Cu K_α) and the divergence of the beam at sample is 0.005^0. The small beam divergence is critical to the analysis of the epitaxial samples. Typical focused X-ray beam flux and beam size for 8keV are 5.6×10^{10} photons/s and 3.1×0.8 mm^2 respectively. The Huber four circle diffractometer located within the end station can perform grazing incidence X-ray reflectivity, high resolution X-ray diffraction and triple axis reciprocal space mapping. Analysis of X-ray data was performed using the software LEPTOS (Bruker). Tensile strain in the ε-Si samples was measured independently by a Jobin Yvon TL64000 UV Raman spectrometer with a 325nm laser source in a backscattering geometry. This technique will be discussed in detail in chapter 5.

4. APPLICATION EXAMPLE: BIAXIAL STRESSED SILICON SUBSTRATES

This method is first applied to biaxial stressed silicon. The optical model used for SE data fitting consists of four layers namely native oxide, surface roughness, ε-Si and relaxed $Si_{1-x}Ge_x$ capping layer (Fig. **3**). The graded $Si_{1-x}Ge_x$ buffer layer and the (001) Si substrate of the samples are not included because even at 500nm, the optical penetration depth in a $Si_{0.805}Ge_{0.195}$ capping layer is only 248nm [17]. This corresponds to an absorption coefficient of 4.03×10^4 cm^{-1}. Since the dielectric function spectrum is of primary interest in this study, a two-stage fitting procedure has been developed to fit the SE data. This was found to be necessary because of the large number of fitting parameters involved.

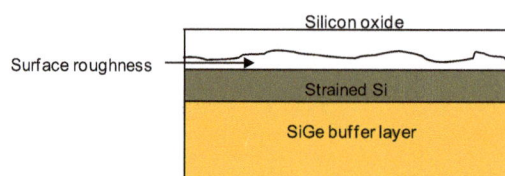

Figure 3: Optical model used for SE data fitting of ε-Si samples.

In the first stage, point-by-point fitting with literature optical constants for bulk Si was used to obtain an initial estimate of the dielectric function of the ε-Si layer. The only other parameter that is allowed to vary is the thickness of the native oxide layer. The ε-Si layer thickness, Ge composition x and the silicon oxide dielectric function were fixed at values as determined from X-ray characterization. Although point-by-point fitting provides a convenient way to obtain the dielectric function given the film thickness, any noise present in the original data can be incorporated into the fitted dielectric function. Thus, the resulting ε-Si dielectric function may not be Kramers-Kronig (K-K) consistent.

In order to ensure that the fitted dielectric function of ε-Si is K-K consistent, a parametric semiconductor (PSM) dielectric function model [18] was used to refine the initial fit. During the second stage fitting, the ε-Si dielectric function obtained from the initial fit was used to find initial values of the parameters in the PSM model. The PSM model was developed by Herzinger and Johs in 1998 to overcome several difficulties inherent to earlier dielectric function models for semiconductor materials [18]. In this model, the complex dielectric function $\varepsilon(\omega)$ is represented as the sum of m polynomial functions and a further P poles may be deployed to account for index effects due to Restrahlen absorption outside the fitted spectral region as shown in (3.10). Such terms however are not used in the present fitting.

$$\varepsilon(\omega) = 1 + \sum_{j=1}^{m} \int_{E_{\min}}^{E_{\max}} W_j(E) \times \Phi(\hbar\omega, E, \sigma_j) dE + \sum_{j=m+1}^{m+1+P} \frac{A_j}{(\hbar\omega)^2 - E_j^2} \tag{3.10}$$

Each of the second summation term of (3.10) consists of a bounded integral of the product of an absorption spectrum $W_j(E)$ and a Gaussian broadening function $\Phi(\hbar\omega, E, \sigma_j)$. The absorption spectrum $W_j(E)$ is defined by the bounded polynomial:

$$W_j(E) = \sum_{k=0}^{N} P_{j,k} E^k u(E - a_j) u(b_j - E) \tag{3.11}$$

where $u(x)$ is the unit step function and a_j and b_j are the lower and upper bounds of the polynomial function respectively. $P_{j,k}$ is the jth coefficient of E^k in the jth polynomial. The Gaussian broadening function Φ is given by (3.12):

$$\Phi(\hbar\omega, E, \sigma) = \sqrt{\frac{\pi}{8\sigma^2}} \left[e^{-y_1^2} + e^{-y_1^2} erf(iy_1) - e^{-y_2^2} - e^{-y_2^2} erf(iy_2) \right] \tag{3.12}$$

$$y_1 = \frac{\hbar\omega - E}{2\sqrt{2}\sigma} \quad \text{and} \quad y_2 = \frac{\hbar\omega + E}{2\sqrt{2}\sigma} \tag{3.13}$$

where σ is the Gaussian broadening parameter and erf is the error function. The expressions (3.13) allow one dimensional lookup tables to be constructed for each order of polynomial so that computation of (3.12) can be facilitated.

For the present samples, the dielectric function of the ε-Si layer was represented by two or three PSM oscillators for the region in the vicinity of the critical points E_1 and E_2 and a further dummy oscillator was used to model the region between the two CPs. Since the PSM model has a large number of parameters, the fitting parameters were first restricted to the native oxide thickness, the roughness layer thickness and the dielectric functions of ε-Si. The ε-Si thickness, Ge composition and SiGe optical constants were fixed [18, 19]. In subsequent iterations, these parameters were made adjustable to arrive at the best fit parameters. For both stages of fitting, the criterion used by the regression algorithm is the mean square error (MSE) defined as:

$$MSE = \frac{1}{2N - M} \sum_{i=1}^{N} \left[\left(\frac{\psi_i^{cal} - \psi_i^{exp}}{\sigma_{exp} \psi_i^{exp}} \right)^2 + \left(\frac{\Delta_i^{cal} - \Delta_i^{exp}}{\sigma_{exp} \Delta_i^{exp}} \right)^2 \right] \tag{3.14}$$

Here, N is the number of data points in the spectrum, M is the number of variable parameters in the optical model, ψ_i^{exp} and ψ_i^{cal} are the experimental and calculated values of ψ and Δ_i^{exp} and Δ_i^{cal} are the experimental and calculated values of Δ respectively. σ_{exp} is the standard deviation of the experimental data points for the quantity concerned.

The best fit dielectric functions of ε-Si layers for S15, S20 and S25 are shown in Fig. 4 together with the dielectric function of bulk Si [12]. The critical points at ~3.4eV and ~4.3eV can be seen clearly in the imaginary dielectric function (ε_2). The edge of E$_1$ shifts to lower photon energy when strain is present and this is consistent with previous reports [12]. A small shift in the peak of E_2 can also be discerned. For the real part of the dielectric function, ε_1 there is a reduction in magnitude near both E_1 and E_2. The E_1 line width broadens when strain is present because the tensile strain splits the hydrostatic shift [1]. It can also be seen that the fitted dielectric function of sample S25 differs significantly from the other two samples.

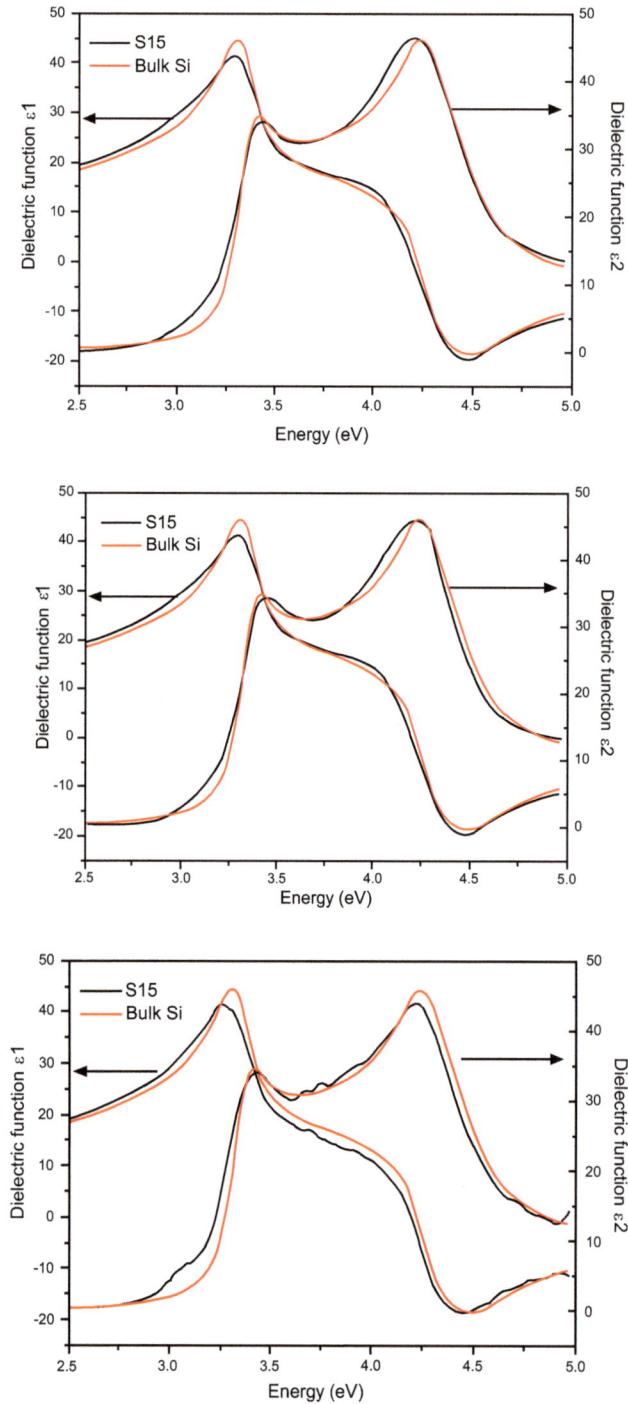

Figure 4: Real part ε_1 and imaginary part ε_2 of the dielectric function of ε-Si layers in samples S15, S20 and S25. The dielectric function of bulk Si is included for comprison. Reprinted with permission 'Characterization of biaxial stressed silicon by spectroscopic ellipsometry and synchrotron X-ray scattering', T.K.S. Wong, Y. Gong, P. Yang, C.M. Ng, Semicond. Sci. and Technol. 22, 1232-1239, November 2007, IOP Publishing. DOI: 10.1088 /0268-1242 /22/11/009.

Since the parameters of the PSM model have little physical meaning, the fitted parameter values cannot be used directly to find the location of the E_1 CP. Instead, line shape analysis has to be performed on the second derivative of the dielectric function spectrum [11]. The second derivative was obtained numerically from the best fit dielectric function of the samples by the Savitzky-Golay algorithm [1]. Sets of consecutive data points in $\varepsilon(\omega)$ were interpolated by a fourth order polynomial using a regression algorithm. The fitted polynomials are then differentiated to yield $d^2\varepsilon/d\omega^2$ as shown in Fig. **5** for sample S15. The second derivative spectra can be described by the Lorentzian line shape:

$$\frac{d^2\varepsilon}{d\omega^2} = n(n-1)Ae^{i\varphi}(\hbar\omega - E + i\Gamma)^{n-2}, \ n \neq 0 \qquad (3.15)$$

where n is the dimension of the CP; A is the amplitude; ϕ is the phase angle; E is the position of the CP and Γ is the peak broadening parameter. As described in ref. [3.11], the E_1 CP in Si has an excitonic character even at room temperature and as a result, $n = -1$.

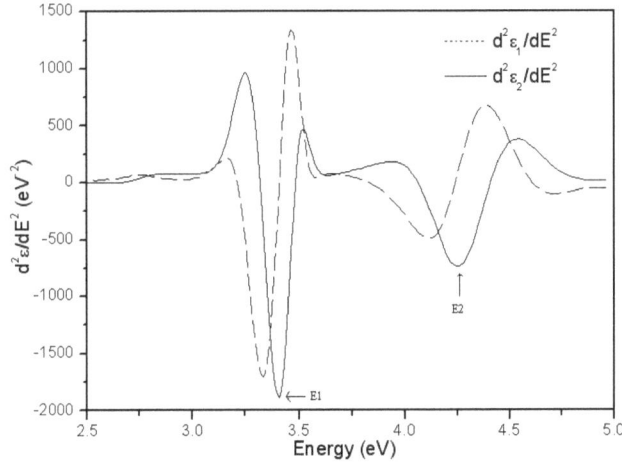

Figure 5: Second derivative of real and imaginary parts of the fitted dielectric function of sample S15. Reprinted with permission 'Characterization of biaxial stressed silicon by spectroscopic ellipsometry and synchrotron X-ray scattering', T.K.S. Wong, Y. Gong, P. Yang, C.M. Ng, Semicond. Sci, and Technol. 22, 1232-1239, November 2007, IOP Publishing DOI: 10.1088/0268-1242/22/11/009.

By using equation (3.13), the critical points E_1 and E_2 of the ε-Si dielectric function can be determined. These are listed together with the shifts in E_1 and E_2 with respect to bulk Si for all three samples in Table **2**. As mentioned above, the E_1 peak shifts to lower energy as the Ge content in the VS increases. For E_2, the peak undergoes a similar shift up to a Ge concentration of 20 at. % and then it saturates. In the following, the shift in E_1 is used to determine the strain in the ε-Si layer. This is due to the lack of deformation potential data for the E_2 peak.

Table 2: Critical Point Shifts for ε-Si Samples

Sample	E_1 (eV)	δE_1 (eV)	E_2 (eV)	δE_2 (eV)
Bulk Si	3.46	0	4.26	0
S15	3.4066	-0.0534	4.2466	-0.0134
S20	3.3787	-0.0813	4.2321	-0.0279
S25	3.3604	-0.0996	4.2321	-0.0279

By substituting equations (3.8) and (3.9) into (3.7), the shift in the position of E_1 can be related to the in plane strain as:

$$\delta E_1 = \frac{2D_1^l}{\sqrt{3}}\left(1-\frac{C_{12}}{C_{11}}\right)\varepsilon_{xx} + \frac{\Delta_1}{2}\left\{1-\left(1+\frac{8}{3}\left(D_3^3\left(1+\frac{2C_{12}}{C_{11}}\right)\right)^2\left(\frac{\varepsilon_{xx}}{\Delta_1}\right)^2\right)^{1/2}\right\} \tag{3.16}$$

With known parameter values, this equation can be solved either analytically or graphically for ε_{xx} for a given δE_1. In these calculations, the parameter values D_1^l = -8.5eV, D_3^3 = -4.3eV and Δ_1 = 44meV were used [11]. For the elastic constants of silicon, the values C_{11} = 16.58x10^{10} Nm^{-2} and C_{12} = 6.39x10^{10} Nm^{-2} are used [15]. The in-plane strain of the samples S15, S20 and S25 are thus determined to be 0.56%, 0.80% and 0.96% respectively (Table **3**). These strains are plotted together with the observed peak shifts in Fig. **1**.

In order to verify this method, validation measurements were performed by two other established methods (Raman and XRD) which will be covered in greater detail in subsequent chapters. Good agreement is found between the SE and Raman deduced strain for samples S15 and S20. This shows that for moderate Ge concentration (up to 20 at. %), the shift of the E_1 peak can serve as a useful qualitative or quantitative indicator of strain. In process monitoring, deviation of the δE_1 shift from known values obtained from reference samples can indicate a change in strain from the designed value. For sample S25, there is greater discrepancy between SE and Raman measurements. It was noted during the SE fitting that the quality of the fit (MSE) for this sample is slightly worse than S15 and S20. Attempts were made to improve the MSE by modifying the default optical model to include additional interfacial layers between the ε-Si and VS. However, the MSE did not improve significantly. This discrepancy is explained in [1].

Table 3: In-Plane Strain of ε-Si Samples

Sample	From SE	From Raman	From XRD	Expected strain
S15	0.56%	0.51%	0.58%	0.56%
S20	0.80%	0.77%	0.76%	0.76%
S25	0.96%	1.03%	1.1%	0.95%

5. SUMMARY

SE has been demonstrated for strain monitoring in ε-Si as well as film thickness measurements. The method is based on the parametric fitting of the dielectric function of ε-Si using supplementary data determined by X-ray scattering. Since SE is readily available, the present method can be suitable for process monitoring. Although a synchrotron source had been utilized for the present work, the present availability of sophisticated X-ray characterization tools should make this approach feasible for semiconductor manufacturing [16]. We have also demonstrated that X-ray metrology is able to reveal subtle structural differences between heteroeptaxial samples.

6. DISCLOSURE

Part of the information included in this chapter has been previously published in Semiconductor Science and Technology Volume 22 Number 11.

REFERENCES

[1]   T.K.S. Wong, Y. Gong, P. Yang and C.M. Ng, "Characterization of biaxial stressed silicon by spectroscopic ellipsometry and synchrotron X-ray scattering," *Semicond. Sci. Technol.*, vol. 22, pp. 1232-1239, Nov. 2007.

[2]   C.J Vineis, M. Erdtmann and C.W. Leitz, "Optimized measurement of strained Si thickness and SiGe virtual substrate composition by spectroscopic ellipsometry," *Thin Solid Films*, vol. 513, pp. 78-13, Aug. 2006.

[3]   O. Fursenko, J. Bauer, P. Zamuseil, D. Kruger, A. Goryachko, Y. Yamamoto, K. Kopke and B. Tillack, "Spectroscopic ellipsometry for in-line process control of SiGe:C HBT technology," *Mater. Sci. Semicond. Proc.*, vol. 8, pp. 273-278, Feb. 2005.

[4]　J. Schmidt, G. Vogg, F. Bensch, S. Kreuzer, P. Ramm, S. Zollner, R. Liu and P. Wennekers, "Spectroscoic techniques for characterization of high-mobility strained-Si CMOS," *Mater. Sci. Semicond. Proc.*, vol. 8, pp. 267, Feb. 2005.

[5]　M.L. Lee, E.A. Fitzgerald, M.T. Bulsara, M.T. Currie and A. Lochtefeld, "Strained Si, SiGe and Ge channels for high mobility metal-oxide-semiconductor field-effect transistosr." *J. Appl. Phys.*, vol. 97, 011101-1-27, Jan. 2005.

[6]　Y. Sun, S.E. Thompson and T. Nishida, "Physics of strain effects in semiconductors and metal-oxide-semicondutor field-effect transistors," *J. Appl. Phys.*, vol. 101, pp. 104503-1-22, May. 2007.

[7]　V. Vartanian, M. Sadaka, S. Zollner, A.V.Y. Thean, T. White, B.Y. Nguyen, M. Zavala, L. McCormick, L. Prabhu, D. Eades, S. Parsons, H. Collard, K. Kim, J. Jiang, V. Dhandapani, J. Hildreth, R. Powers, G. Spencer, N. Ramani, J. Mogab, M. Kottke, M. Canonico, Q. Xie, X.D. Wang, J. Vella, L. Contreras, D. Theodore, B. Lu, T. Kriske, R. Gregory and R. Liu, "Metrology challenges for 45nm strained-Si devices," In: *Characterization and metrology for ULSI Technology*, 2005 pp. 214-221.

[8]　H.J. Grossman, B.A. Davidson, G.J. Gualtieri, G.P. Schwartz, A.T. Macrander, S.E. Slusky, M.H. Grabow and W.A. Sunder, "Critical layer thickness and strain relaxation measurements in GaSb(001)/AlSb structures," *J. Appl. Phys.*, vol. 66, pp. 1687-1694, Aug. 1989.

[9]　H.G. Tompkins and W.A. McGahan, *Spectroscopic Ellipsometry and Reflectometry A Users Guide,* Wiley: New York, 1999.

[10]　P. Etchegion, J. Kircher and M. Cardona, "Elasto-optical constants of Si," *Phys. Rev. B*, vol. 47, pp. 10292-10303, Apr. 1993.

[11]　P. Lautenschlager, M. Garriga, L. Vina and M. Cardona, "Temperature dependence of the dielectric function and interband critical points *in silico*n," *Phys. Rev. B,* vol. 36, pp. 4821-4830, Sep. 1987.

[12]　H. Lee and E.D. Jones, "Dielectric function of biaxially strained silicon layer," *Appl. Phys. Lett.*, vol. 68, pp. 3153-3155, May. 1996.

[13]　R. Hull, Ed., *Properties of Crystalline Silicon,* INSPEC: London, 1997.

[14]　P. Etchegoin, J. Kicher, M. Cardona and C. Grein, "Piezo-optical response of Ge in the visible-uv range," *Phys. Rev. B,* vol. 45, pp. 11721-11735, May. 1992.

[15]　O. Madeulung, M. Schultz and H. Weiss, *Landbolt-Bornstein Semiconductors: Group IV Eelements and III-V Ccompound,* Springer: Berlin, 1982.

[16]　S. Zollner, J.P. Liu, P. Zaumseil, H.J. Osten and A.A. Demkov, "Dielectric functions, elasto-optic effects, and critical-point parameters of biaxialy stressed $Si_{1-y}C_y$ alloys on Si (001)," *Semicond. Sci. Technol.*, vol. 22, pp. S13-S20, Jan. 2007.

[17]　C. Pickering and R.T. Carline, "Dielectric function spectra of strained and relaxed $Si_{1-x}Ge_x$ alloys (x = 0 – 0.25)," *J. Appl. Phys.,* vol. 75, pp. 4642-4647, May. 1994.

[18]　B. Johs, C.M. Herzinger, J.H. Dinan, A. Cornfeld and J.D. Benson, "Development of a parametric optical constant model for $Hg_{1-x}Cd_xTe$ for control of composition by spectroscopic ellipsometry during MBE growth," *Thin Solid Films*, vol. 313-314, pp. 137-142, Feb. 1998.

[19]　P. Lautenschlager, M. Garriga, L. Vina and M. Cardona, "Temperature dependence of the dielectric function and interband critical points in silicon." Phys. Rev. B, vol. 36, pp. 4821-4830, Sep. 1992.

CHAPTER 4

Photoreflectance Method

Abstract: A modulated spectroscopy method called photoreflectance is described. The technique involves using an incident beam and a coincident modulating beam usually from a diode laser. The former results in a reflectance signal while the latter gives a modulated reflectance signal. By fitting the spectrum of the normalized modulated reflectance with a third order derivative of the dielectric function, the critical point energies can be determined with high accuracy. Two photoreflectance spectroscopy instruments with high throughput are discussed. The use of these instruments is illustrated by examples from biaxially stressed silicon on insulator substrates and thin film photovoltaic devices. A recent method to suppress spurious modulated light from entering the photodetector is also briefly discussed.

Keywords: Photoreflectance, Modulated spectroscopy, Critical point, Strained silicon, Photovoltaics.

1. INTRODUCTION AND BACKGROUND

Photoreflectance spectroscopy is a non-destructive optical characterization technique that has recently been applied to measure strain in biaxially stressed silicon substrates. This technique can be deployed in the semiconductor fabrication facilities as an in-line process monitoring tool because the measurement can be carried out relatively quickly through parallel acquisition of the photoreflectance spectrum. A rapid photoreflectance spectroscopy instrument has been demonstrated by Optical metrology innovations Ltd. (Cork, Ireland) in 2008 [1]. Another company Xitronix (Austin, Texas) has developed a photoreflectance characterization tool specifically for the semiconductor fabrication industry. One of the main applications of this technique is to evaluate strain in strained silicon-on-insulator (SSOI) substrates and this example will be described later in this chapter. However, at present, this technique lacks the spatial resolution for probing strain within individual devices. Thus, it cannot be used for characterizing process induced strain and in this respect is similar to the spectroscopic ellipsometry technique in the previous chapter.

The photoreflectance technique is one of a family of modulated spectroscopy techniques and its origin dated back to 1964 [2]. In the literature, there is a very detailed review of this field up to 1993 by one of the pioneers of this field [3]. In modulation spectroscopy, the optical spectral response of the sample is studied applying a periodic perturbation such as an electric field, heat pulse or external stress. These are referred as electromodulation, thermomodulation and piezomodulation respectively. For a long time, the electromodulation technique was used only for fundamental studies of semiconductor band structure and it is still used for this purpose in nanostructure research [2, 4]. As will be shown shortly, the method is highly sensitive to the critical point transition energies. This is because the spectra obtained are of a derivative-type and thus they peak strongly near all critical points. Furthermore, the spectra can be acquired at room temperature and unlike luminescence, does not require cryogenic conditions. As with SE, it can provide accurate information on these energies. From a metrology standpoint, this will enable the strain to be determined.

2. MEASUREMENT PRINCIPLE

Photoreflectance spectroscopy as the name implies is based on the measurement of reflected light as a function of the wavelength or photon energy of the incident light. It is a form of non-contact electromodulated reflectance experiment [2]. The alternative method of introducing electric field modulation is to deposit contacts on the top and bottom or on the surface of the sample and apply a voltage. However, this requires the sample to be processed into device like test structures.

When a light beam is incident at a reflective semiconductor surface at an angle, a portion of its energy may be reflected specularly from the air-substrate interface because of the difference in dielectric constants or permittivity of the two media. The reflectance is simply the magnitude ratio of the energy of the reflected beam to the energy of the incident beam. The reflectance is also called the reflectivity and the energy in the incident and reflected beam is given by the square of the wave amplitude.

Terence K.S. Wong

In a photoreflectance experiment, a second beam called the pump beam is used to apply energy to the spot on the sample where the reflection occurs. The wavelength of the pump beam is chosen such that it can be absorbed by the semiconductor. In practice, this means the energy of the photons in the pump beam is greater than the bandgap energy of the semiconductor. As a result of this, absorption of the pump beam will occur near the surface of the semiconductor and this will generate excess electron hole pairs. The presence of these charge carriers will alter the optical properties of the semiconductor such as the complex permittivity and so whenever the pump beam is applied, there will be a change in the reflectance which is usually denoted as ΔR where R is the reflectance in the absence of the pump beam. The photoreflectance spectroscopy technique is basically about measuring the two quantities, R and ΔR, as a function of the photon energy. From a measured photoreflectance spectrum, one can optically probe the electronic band structure of a semiconductor near the band maxima and minima. This is, in fact, the main application of this technique in the past. An example of experimental photoreflectance spectra is shown in Fig. **1**.

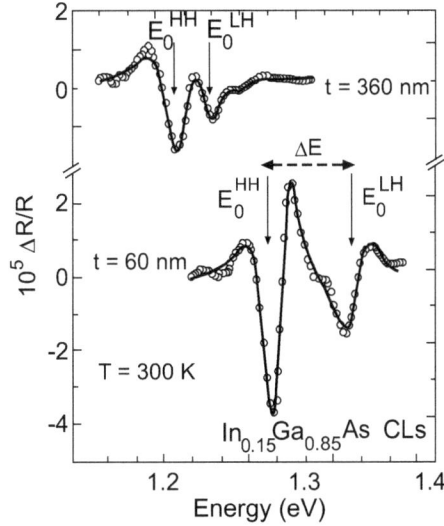

Figure 1: Example of a room temperature photoreflectance spectrum from a semiconductor nanostructure [4]. Reprinted with permission from M. Geddo, G. Guizzetti, M. Patrini, T. Ciabattoni, L. Seravalli, P. Frigeri and S. Franchi, "Metamorphic buffers and optical measurements of residual strain", *Appl. Phys. Lett.,* vol. 87, pp. 263120-1-3, 2005. Copyright 2005, American Institute of Physics.

In the following, we provide a more in-depth discussion of the photoreflectance technique in the context of strain characterization. The reflectance R of a semiconductor surface is determined by the real and imaginary parts of the complex refractive index of the semiconductor, $n+ik$, where n is the refractive index and k is the extinction coefficient. Both n and k are functions of photon energy and they are the basic optical properties of a semiconductor material. One can square the complex refractive index and obtain the complex permittivity which is an alternative way of representing the optical properties of the semiconductor. The complex permittivity has a real part ε_1 and a complex part ε_2. If the complex refractive index is used to find the reflectance, then application of electromagnetic theory will give:

$$R = \frac{(n-1)^2 + k^2}{(n+1)^2 + k^2} \tag{4.1}$$

The extinction coefficient k is related to the absorption coefficient α of the semiconductor by $\alpha = 4\pi k / \lambda$ where λ is the free space wavelength. When the incident wavelength is longer than the absorption edge of the semiconductor, the absorption will be quite negligible and the reflectance can be simplified to:

$$R = \frac{(n-1)^2}{(n+1)^2} \tag{4.2}$$

This shows somewhat more clearly that if there were to be a prominent change in the value of n at some photon energy or wavelength, then this change should be reflected in *R*. The refractive index spectrum or refractive index as a function of photon energy is fundamentally determined by the energy band structure of the semiconductor and thus is characteristic for each semiconductor. The refractive index spectrum can be found mathematically if the extinction coefficient spectrum (or equivalently the absorption spectrum) is known. This is because of the Kramers-Kronig relationship (chapter 2) which states that the refractive index is actually:

$$n(E) = 1 + \frac{ch}{2\pi} P \int_0^\infty \frac{\alpha(E')}{(E')^2 - E^2} dE'$$

(4.3)

where *P* is called the Cauchy integral and *E'* is a dummy integration variable for energy. *c* is the velocity of light and *h* is the Planck constant. If we treat the integrand inside the Cauchy integral as a product of two functions of *E'* and apply integration by parts, then the refractive index will be:

$$n(E) = 1 + \frac{ch}{4\pi} P \int_0^\infty \ln\frac{1}{(E')^2 - E^2} \frac{d\alpha(E')}{dE'} dE'$$

(4.4)

The Kramers-Kronig relationship is rewritten like this because it emphasizes that whenever the absorption coefficient is changing rapidly with energy, there should be a change in the refractive index as well. This correspondence can be seen easily in any ellipsometry experiment. The absorption coefficient will increase when the energy of the incident photons matches the energy difference between a pair of critical points in the energy band structure of the semiconductor. The concept of critical points had already been discussed in chapter 2 and for clarity, we will reiterate its meaning here. The critical points of a semiconductor band structure refer to any points in momentum space that satisfies the condition, $\nabla_k E_k = 0$. Since in Cartesian coordinates, the components of the gradient of E_k is simply the derivatives of E with respect to k_x or k_y or k_z, the condition implies the existence of four types of critical points, namely, minimum point M_0, saddle points M_1, M_2 and maximum point M_3. An optical transition such as absorption always involves a pair of critical points and thus we can also write

$$\nabla_k \left(E_c(k) - E_v(k) \right) = 0$$

(4.5)

Here E_c is the conduction band structure and E_v is the valence band structure. For each pair of electronic states in the conduction and valence band with energy difference *E*, there is a joint density of states (JDOS) function which had also been discussed in chapter 2. The JDOS function gives the number of electronic states per unit volume per unit energy interval that can participate in an optical transition and determines the strength of that transition. Thus, when the JDOS increases, the absorption coefficient α also tends to become large. From the Kramers-Kronig relation above (4.4), there should also be a sharp change in the refractive index *n* near any optical transitions between a pair of critical points. Since the reflectance depends on both *n* and *k*, the reflectance must show a signature feature in its spectrum as well and this is what one tries to measure in a photoreflectance experiment.

For many semiconductors, the change in the reflectance of the sample near a critical point transition is not as large as one would like. Thus, it is often necessary to enhance the peaks in the reflectance spectrum by using a modulation technique to allow a more accurate determination of the energy of the transitions. By modulation, we mean an excitation is applied periodically to the sample to affect the JDOS. The most common excitation is an electric field. It can be shown that when an electric field, E is present, the change in refractive index is given by [2]:

$$\Delta n(E, \mathrm{E}) = \frac{ch}{\pi} P \int_0^\infty \frac{\Delta\alpha(E', \mathrm{E})}{(E')^2 - E^2} dE'$$

(4.6)

Note that this is in fact an application of the Franz-Keldysh effect which states that the absorption spectrum of a semiconductor is shifted by a strong electric field [2]. The simplest way to apply a modulating electric field is of course to use a pair of electrodes on the semiconductor. This is in fact often done for research purposes and is called an electroreflectance experiment. However, for the purpose of rapid strain characterization, the deposition of electrodes onto the semiconductor is not feasible and an optical modulation of the electric field is used instead. From basic semiconductor theory, we know that the surface of any unpassivated semiconductor is filled with surface electronic states. These states arise from the abrupt termination of the semiconductor crystal and the resulting dangling bonds are what prevented the MOSFET from being developed for a long time in the twentieth century. Due to these surface states, there is band bending at the surface and this causes a static electric field at the semiconductor surface. Thus, an electric field perpendicular to the surface is present at the near surface region of the semiconductor. The magnitude of this electric field can be reduced to zero temporarily by using a large concentration of charge carriers generated optically at the near surface region by a laser pulse that is strongly absorbed by the sample. If a pulsed laser is used, then the surface electric field and the refractive index can be modulated. This change in the refractive index can be detected accurately by a synchronous or lock-in technique. By using the spectrum of $\Delta R/R$, one can obtain a more precise measurement of the energy of the optical transition. This photoreflectance technique is actually not as sensitive as the electroreflectance technique but it has the advantage of being non-intrusive and requires no further sample preparation. Hence the method is used for strain characterization.

Why is the normalized modulated reflectance $\Delta R/R$ used in photoreflectance spectroscopy? The reason is that at the low electric field regime, the normalized change in reflectivity is actually related to the change in the dielectric function $\Delta \varepsilon$ by the Seraphin equation [3]:

$$\frac{\Delta R}{R} = a\Delta\varepsilon_1 + b\Delta\varepsilon_2 \tag{4.7}$$

Here, a and b are the Seraphin coefficients and $\Delta\varepsilon_1$ and $\Delta\varepsilon_2$ are the changes in the real and imaginary parts of the dielectric function. If absorption is not significant, then the Seraphin coefficient b can be neglected and the normalized modulated reflectance is simply proportional to the change in the real part of the dielectric function. The shape of this function according to the pioneer of this field, D.E. Aspnes is the third derivative of the unperturbed dielectric function ε. Aspnes further proposed that a third derivative function form (TDFF) should be used to fit the data from photoreflectance spectroscopy [5].

$$\frac{\Delta R}{R}(E) = \mathrm{Re}\left[A\exp\left(i\theta\right)\left(E - E_g + i\Gamma\right)^{-m} \right] \tag{4.8}$$

Here A is an amplitude, θ is a phase factor, E_g is the energy bandgap or the critical point in the joint density of states function and Γ is the line broadening. The parameter m typically has a value between 2.5 and 3.5 depending on the dimension of the critical point. Re stands for the real part of the function inside the brackets. In calculus, any differentiation and subsequent differentiation will accentuate any turning point and smooth variations will be reduced to zero. The reader should now realize why photoreflectance spectroscopy is highly sensitive to the critical point features in the dielectric function and why it is suitable for strain characterization.

The form of the TDFF will be shown in the following section. However, it is pointed out that each critical point is situated in between a local maximum and minimum in the $\Delta R/R$ spectrum. The TDFF as shown in (4.8) applies only to low field $\Delta R/R$ spectra.

3. EXPERIMENT AND INSTRUMENTATION

Photoreflectance experiments had been used to study semiconductor band structure for a long time. As pointed out in the introduction, there are numerous references on this technique to study band structure effects in semiconductors [3]. However, in these experiments, the photoreflectance spectrum ($\Delta R/R$) is typically measured serially or one wavelength at a time and the whole spectrum can take considerable time to acquire. Therefore this type of photoreflectance measurement is not suitable for in-line semiconductor metrology.

In this section, we first provide a description of a rapid photoreflectance spectroscopy (RPR) technique that had been developed by Chouaib and co-workers [1]. A schematic diagram of this instrument is shown in Fig **2**. The modular instrument consists of an incident light source from a Xenon arc lamp and the probing light from this source is directed to the semiconductor sample through an optical fibre. The reflected light from the sample is filtered to give a range of wavelengths near the spectral region of interest. In the case of strained silicon, this would be near the *E1* critical point at 3.4eV. The reflected light is dispersed into different wavelengths by a spectrograph. This is an optical instrument in which light propagates at different angles with respect to the incident polychromatic light depending on the wavelength. At the exit end of the spectrograph, there is an array of photodetectors. Each photodetector is capable of rapidly detecting a narrow band of wavelengths and so the array as a whole can detect the entire wavelength range of the probe beam simultaneously.

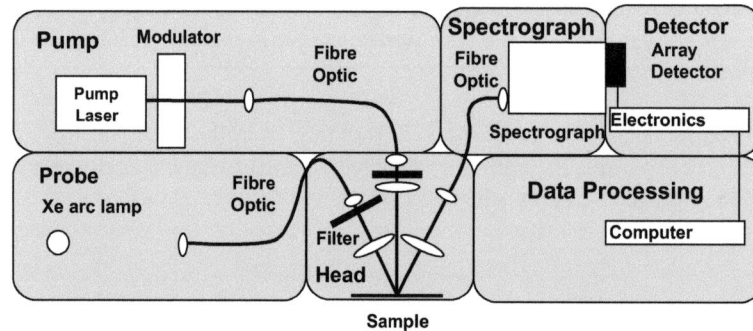

Figure 2: Schematic diagram of the rapid photoreflectance spectroscopy apparatus for the measurement of strained semiconductors such as strained silicon [1]. Reprinted with permission from H. Chouaib, M.E. Murtagh, V. Guenebaut, S. Ward, P.V. Kelly, M. Kennard, Y.M. le Vailant, M.G. Somekhm M.C. Pitter and S.D. Sharples, "Rapid photoreflectance spectroscopy for strained silicon metrology", *Rev. Sci. Instrum.,* vol. 79, pp. 1031061-3, 2008. Copyright 2008, American Institute of Physics.

The pump beam which generates the modulated reflectance comes from a laser. The laser light is modulated in intensity by a rotating chopper wheel and the light is also coupled to the sample through an optical fibre. The modulated laser light illuminates the sample at the same spot as the probe beam and as a result, there is a periodic change in the reflectance ΔR. This change in reflectance is typically much smaller than the static reflectance at a given wavelength. However, since it is a periodic signal, one can use phase sensitive lock-in amplifiers to selectively detect this weak signal component.

The RPR technique is rapid because the electronics is fast enough to read all the detectors in the array by a multiplexing technique. For each detector, multiple measurements of ΔR and R are made during each modulation period. As a result, the total time for measuring one RPR spectrum is only about 1ms [1]. In practice, the same spectrum is acquired several thousand times per sample and all the spectra are averaged to obtain a good signal to noise ratio.

The photoreflectance instrument built by Xitronix is also capable of high throughput [6]. However, it uses a different single wavelength principle to characterize strain. In this instrument, the probe beam is from a monochromatic laser with a wavelength of 375nm. This wavelength is chosen because it is close to the direct interband (*E1*) transition in silicon. When the probe beam is close to either side of the onset of absorption, the sign of the $\Delta R/R$ signal becomes very sensitive to the phase of the pump beam; *i.e.* as the pump phase is varied, the photoreflectance signal will change from positive to negative or *vice versa*. Thus, this instrument does not scan the photoreflectance spectrum on either side of the transition but instead use the sensitivity of the photoreflectance signal to the phase of the pump beam to give an indication of strain in the sample. Since this is effectively a single wavelength measurement, rapid measurement is possible. In their publication [6], the measurement time is reported to be 1 second or less which is much faster than the target set in the 2007 ITRS roadmap for this type of metrology.

3.1. Spurious Light Suppression

A common experimental problem in photoreflectance spectroscopy is that the rather weak ΔR signal is swamped by another (modulated) spurious signal that makes the normalization difficult and yield unreliable results. The cause of this problem is that the light entering the photodetector consists of not just the static reflected light from the probe beam and the modulated reflected component. In addition, there is a spurious third component of scattered light from the pump beam and possibly photoluminescence excited by this beam that is also modulated by the mechanical chopper [2]. Since this spurious light component has the same frequency as the ΔR signal, the phase sensitive detecting device such as a lock in amplifier will treat it as a proper signal and output a recovered signal that has a large offset error. It is therefore important to devise ways to suppress this spurious light and preventing it from affecting the photoreflectance spectra.

Many spurious light suppression schemes have been proposed in the literature [7]. These include dual chopped photoreflectance, sweeping photoreflectance, electrical front end compensation and the use of a Fourier transform spectrometer. However, they all have their disadvantages and add complexity to the experimental setup. A simple, elegant method called optical front-end compensation (OFEC) or spurious light suppression was recently demonstrated by researchers at the Australian National University in Canberra [7]. The basic idea is to use a second compensation beam that is equal in amplitude but opposite in phase to the first pump beam to suppress the spurious light reaching the photodetector (Fig. **3**). Both the pump beam and compensation beam use the same mechanical chopper but the beams pass through different blades of the wheel so that a desired phase difference can be set using a movable mirror (Fig. **4**). The intensity of the two beams is matched by using a variable neutral density filter. An important part of this technique is to position the compensation beam at a spot as close as possible but not coincident with the pump beam. This is because obviously, the compensation beam may then affect the modulated reflectance signal. If this placement of the compensation beam is done carefully, then the spurious light from the pump beam will be almost cancelled by the compensation beam in antiphase and the intensity of the spurious light will be greatly reduced.

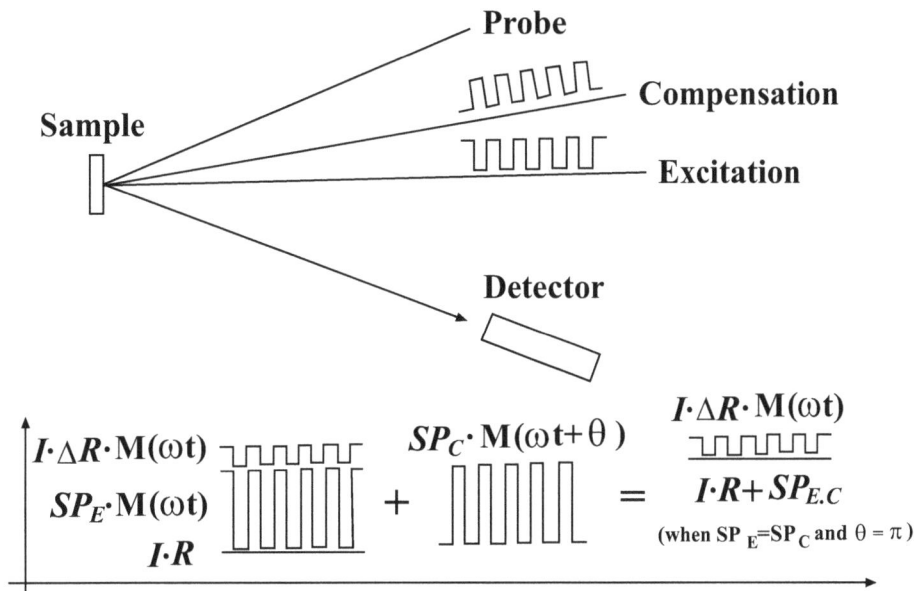

Figure 3: Principle of the optical front end compensation technique for suppression of spurious light in a photoreflectance experiment [7]. Reprinted with permission from Q. Li, H.H. Tan and C. Jagadish, "A new optical front-end compensation technique for suppression of spurious signal in photoreflectance spectroscopy using an antiphase signal", *Rev. Sci. Instrum.*, vol. 81, pp. 0431021-4, 2010. Copyright 2010. American Institute of Physics.

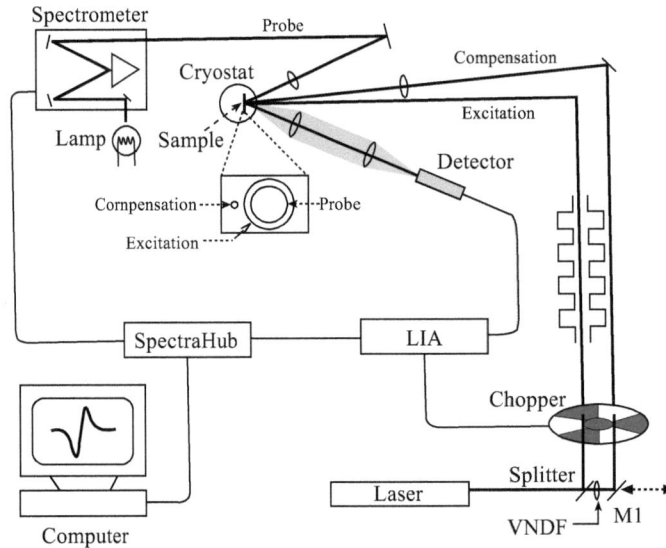

Figure 4: Experimental setup of a photoreflectance spectrometer incorporating optical front end compensation [7]. Reprinted with permission from Q. Li, H.H. Tan and C. Jagadish, "A new optical front-end compensation technique for suppression of spurious signal in photoreflectance spectroscopy using an antiphase signal", *Rev. Sci. Instrum.*, vol. 81, pp. 0431021-4, 2010. Copyright 2010. American Institute of Physics.

4. APPLICATION EXAMPLES

Several examples will be given in this section to illustrate how the photoreflectance spectroscopy technique has been used to measure strain in semiconductor samples. The examples are not limited to strained silicon layers even though this material will be covered first. Due to the growing importance of photovoltaic devices, we will discuss one example of strain effects in these devices by photoreflectance spectroscopy.

4.1. Strained Silicon

By using the RPR spectroscopy measurement system described in section 4.3, Chouaib *et al.* [1] acquired the photoreflectance spectrum of a strained silicon on insulator (SSOI) wafer from 2.7eV to 3.6eV. This energy range spans the E1 transition in silicon. The SSOI wafer was manufactured by SOITEC using the SmartCut^{TM} process [1] and the strained silicon layer was 20nm thick. The entire photoreflectance spectrum consisting of 512 pixels was acquired in 10s.

As with the spectroscopic ellipsometric spectra in chapter 3, the RPR spectrum consists of a series of spectral 'lines' with a finite line width. Each spectral line is manifested as a maximum and a minimum and the entire spectrum needs to be fitted mathematically to find the positions of the critical points to measure the energy of the interband transitions.

In their work, Chouaib *et al.* used two of the following third derivative functional forms to fit the RPR line shapes. This type of function was first proposed by D.E. Aspnes in 1973 and had been defined in section 2.

$$\frac{\Delta R}{R}_{PR}(E) = \mathrm{Re}\left[A\exp\left(i\theta\right)\left(E - E_g + i\Gamma\right)^{-3} \right] \qquad (4.9)$$

After mathematical fitting, two energy gaps were found at 3.346eV and 3.476eV. The first is associated with a light hole (LH) valence band and the latter with the heavy hole (HH) valence band. These two band gaps arise from the splitting of the direct band gap at 3.4eV of the unstrained silicon by the applied strain. As a result of the compressive strain in the normal direction of the wafer [1], the double degeneracy at the top of the valence band is lifted and this results in two slightly different energy band gaps.

Since the energy shifts of the E1 transition at 3.4eV in silicon is due to strain, there should be an equation relating the two. This relation was found by Kondo and Moritani in 1976 using the electroreflectance data from silicon under uniaxial stress [1].

$$\Delta E_{HH} = E_{HH} - 3.4 = \sqrt{\frac{1}{3}} D_1^1 \left(\varepsilon_n + 2\varepsilon_p \right) + \sqrt{\frac{2}{3}} D_3^3 \left(\varepsilon_n - \varepsilon_p \right) \tag{4.10}$$

$$\Delta E_{LH} = E_{LH} - 3.4 = \sqrt{\frac{1}{3}} D_1^1 \left(\varepsilon_n + 2\varepsilon_p \right) - \sqrt{\frac{2}{3}} D_3^3 \left(\varepsilon_n - \varepsilon_p \right)$$

In these two equations, D_1^1 is the hydrostatic deformation potential and has the value of -9.8eV according to Kondo and Moritani. D_3^3 is the intraband strain deformation parameter in the direction perpendicular to the plane of the wafer and has a measured value of 4.7eV. ε_n is the strain in the normal (perpendicular) direction and ε_p is the strain in the plane of the wafer. E_{HH} and E_{LH} are the energy of the heavy hole and light hole transitions in the strained silicon. By using these values, the energy shifts of the *E1* transition can be written more simply as:

$$\Delta E_{HH} = -5.658 \left(\varepsilon_n + 2\varepsilon_p \right) + 3.837 \left(\varepsilon_n - \varepsilon_p \right) \tag{4.11}$$

$$\Delta E_{LH} = -5.658 \left(\varepsilon_n + 2\varepsilon_p \right) - 3.837 \left(\varepsilon_n - \varepsilon_p \right)$$

The in plane strain and normal strain are related by the elastic constants (or Poisson's ratio) of silicon *via* the relationship: $\varepsilon_n = -2 \left(C_{12} / C_{11} \right) \varepsilon_p$. Hence, one can simplify the above equations further and find:

$$\Delta E_{LH} = -0.1375 \varepsilon_p \tag{4.12}$$

This equation shows that the in-plane strain can be calculated from the shift of the light hole transition relative to the *E1* transition in unstrained silicon. This in-plane strain is usually what is needed for strained silicon wafer monitoring. In their paper, Chouaib also checked the in-plane strain measured by RPR against strain measured by Raman spectroscopy and found that the agreement is good and consistent with the experimental uncertainty of the deformation potentials.

For the instrument by Xitronix, only a limited amount of experimental data was reported in the literature [6]. In one proof of principle experiment, two sets of strained silicon wafers were prepared. In the first set, there are two strained silicon wafers with a 6nm strained silicon layer on relaxed silicon germanium (18.5 at.% germanium); an unstrained silicon control wafer and two relaxed silicon germanium wafers with 18.5 at.% germanium. The second set consists of eight strained silicon on silicon germanium wafers with various silicon thickness and germanium concentration.

For each wafer in the first set, 100 measurements were taken by the photoreflectance instrument along a length of 200 microns. All the unstrained substrates show a small negative $\Delta R/R$ signal of $\sim$ -10^{-4} to -10^{-3}. Both the strained silicon wafers, however, showed a positive $\Delta R/R$ signal of $\sim 10^{-3}$. These results showed that a negative signal is correlated with the absence of strain and a positive signal indicates presence of strain.

For the wafers in the second set, five of the wafers showed negative $\Delta R/R$ signals while the other three show positive signals. These three wafers are therefore still strained with a strained silicon thickness of 10nm or less. The other wafers have relaxed presumably because the silicon thickness (20nm) has exceeded the critical thickness which should be between 10nm – 20nm. In addition to strain this instrument can also be used to characterize dopant activation in ultra shallow junctions [6]. This is because the photoreflectance signal is sensitive to the surface crystallinity of the sample.

Strain in ultrathin SSOI wafers were measured by Mungia and collaborators in France using three optical methods including: photoreflectance, Raman spectroscopy and low temperature photoluminescence [8]. The strained silicon was grown on a relaxed $Si_{0.8}Ge_{0.2}$ metamorphic substrate. It was then transferred to a silicon wafer with the Smartcut^{TM} process. For the photoreflectance measurement at room temperature, a frequency doubled 244nm argon laser line was used as the pump beam. The probe beam was from a halogen lamp and a monochromator and the reflectance was measured by a GaAs photomultiplier. The photoreflectance spectrum from 3.0eV to 3.8eV had a number of peaks and valleys and can only be fitted properly by three transition energies using the Aspnes model. From the data fitting, the interband transition energies are at: 3.19eV, 3.30eV and 3.45eV. These correspond respectively to a lower conduction band to light hole band transition; lower conduction band to heavy hole transition and higher conduction band to light hole band transition. Using these transition energies for the strained silicon, the energy band gap shrinkage can be found. The deformation potential is then applied to find the tensile strain of ~1%. This strain is consistent with the values obtained by the other two methods.

In a more recent study [9], this same French group studied three SSOI (sSi20, sSi30, sSi40) wafers with different amount of biaxial strain. The strain was varied by using the germanium composition in the buffer layer of the starting wafer. The three SSOI wafers were by photoreflectance spectroscopy at room temperature as well as by Raman. For all three SSOI wafers, the $\Delta R/R$ spectrum has a more complicated structure than bulk silicon and three TDFF were needed for proper fitting (Fig. **5**). As a result, three bandgaps and their dependence on strain could be deduced. The first (direct) transition E_g is between the bottom of the conduction band at k = 0 and the top of the light hole valence band and this could be well fitted by the following theoretical expression [9]:

$$E_g = E_0^{'} + a\left(\varepsilon_n + 2\varepsilon_{//}\right) + \frac{\Delta_0}{2} - 2b\left(\varepsilon_n - \varepsilon_{//}\right) - \frac{1}{2}\sqrt{\Delta_0^2 + 2\Delta_0\left(b\left(\varepsilon_n - \varepsilon_{//}\right) + 9b^2\left(\varepsilon_n - \varepsilon_{//}\right)^2\right)} \qquad (4.13)$$

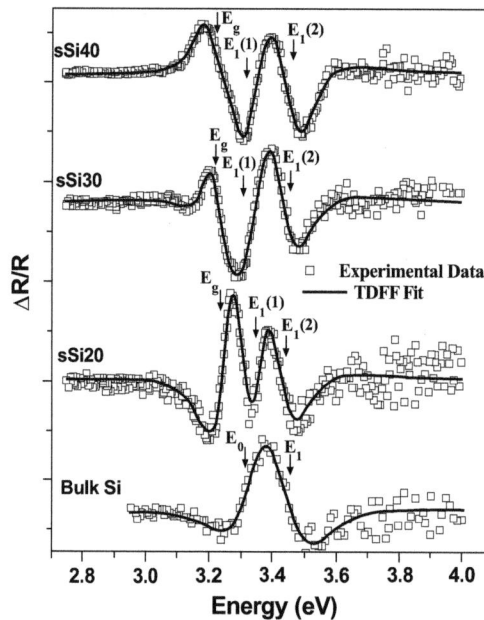

Figure 5: Measured photoreflectance spectrum of three thin strained silicon layers (sSi20, sSi30 and sSi40) [9]. Reprinted with permission from 'Strain dependence of the direct energy bandgap in thin film silicon on insulator layers", J Munguia, J-M Bleut, H Chouaib, G Bremond, M Mermoux and C Bru-Chevallier, J. Phys. D: Appl. Phys. 43, 255401, June 2010, IOP Publishing DOI: 10.1088/0022-3727/43/25/255401

In (4.13), E_g is the bandgap of silicon at the Γ point. $E_0^{'}$ is the direct bandgap of unstrained silicon and is equal to 3.345eV at room temperature. ε_n and $\varepsilon_{//}$ are the strain perpendicular to and in the plane of the silicon. The coefficients, a and b are the bandgap hydrostatic and shear deformation potentials and Δ_0 is the spin orbit coupling.

The other two bandgaps are attributed to *E1* transitions at higher energy in the Λ direction of the Brillouin zone. There are two such gaps, $E_1(1)$ and $E_1(2)$ because of the strain induced splitting of the valence band. Their strain dependence could also be fitted by two theoretical expressions [9].

$$E_1(1) = E_1 + \frac{2D_1^1}{\sqrt{3}}\left(S_{11} + 2S_{12}\right)\frac{E\varepsilon_{//}}{(1-v)} + \sqrt{\frac{2}{3}}D_3^3\left(S_{11} - S_{12}\right)\frac{E\varepsilon_{//}}{(1-v)} \tag{4.14}$$

$$E_1(2) = E_1 + \frac{2D_1^1}{\sqrt{3}}\left(S_{11} + 2S_{12}\right)\frac{E\varepsilon_{//}}{(1-v)} - \sqrt{\frac{2}{3}}D_3^3\left(S_{11} - S_{12}\right)\frac{E\varepsilon_{//}}{(1-v)} \tag{4.15}$$

In (4.14) and (4.15), S_{11} and S_{12} are the compliance components of silicon. D_1^1 and D_3^3 are the hydrostatic and valence band deformation potentials respectively. All four quantities are known as is the Poisson's ratio v. Thus, by measuring the energy gap $E_1(1)$ or $E_1(2)$ of the samples, the in plane strain of the strained silicon can be determined (see [9]).

4.2. Thin Film Photovoltaic Devices

Physical vapor deposited ternary and quartenary (I-III-VI) chalcopyrites semiconductors such as $Cu(In,Ga)S_2$ and $Cu(In,Ga)Se_2$ have important applications as absorber layers in high conversion efficiency thin film photovoltaic devices. The Ga added to a ternary chalcopyrite such as $CuInS_2$, can increase the bandgap from ~1eV towards the ideal value of 1.5eV for photovoltaic applications. The tuning of the bandgap also allows the fabrication of tandem solar cells in which successive absorber layers with different composition are used to absorb different portions of the solar spectrum. For multilayer thin film photovoltaic devices, it is useful to know how strain in a multilayer structure can modify the absorber bandgap. In view of this, a study on the bandgap energies and strain effects in $CuIn_{1-x}Ga_xS_2$ was carried out recently by a group at the National Technical University in Greece [10].

Thin film photovoltaic devices with the layer structure $ZnO/CdS/CuIn_{1-x}Ga_xS_2/Mo/Glass$ ($x = 0$, 0.04, 0.12) were fabricated with Ni/Al top electrodes. The ZnO is a window layer and faces the sunlight. The CdS is a buffer layer deposited by chemical bath method [10]. The $CuIn_{1-x}Ga_xS_2$ (CIGS) layer is deposited by a two stage evaporation process in order to control the concentration of Ga. Since complete solar cells are the samples being investigated, it was possible in this study to perform both contact electroreflectance as well as photoreflectance spectroscopy.

The photoreflectance measurement was carried out at cryogenic temperature (20K) using a 488 nm excitation laser modulated by a mechanical chopper and a closed cycle helium cryostat. The experiment had to be performed at low temperature because at room temperature, the signal to noise ratio was too low for data fitting to be carried out. The photoreflectance spectra at 20K (Fig. **6**) for all three compositions of CIGS were fitted using two TDFF and these suggest two transitions or bandgaps that were referred as E_a and E_b. These transitions were attributed to the lifting of the degenerate states of the valence band maximum in the band structure of the chalcopyrites.

More detailed information about strain effects in these CIGS absorbers were obtained by room temperature electroreflectance. An electroreflectance measurement can be easily performed in this case because contacts were fabricated along with the absorbers on these device samples. The electric field modulation was provided by a 0 - +1.5V square wave from a pulse generator. A more complicated electroreflectance spectrum was obtained in the range 1.3 - 1.8 eV. For the CIGS absorbers with $x = 0.04$ and 0.12, four TDFF functions were required for a proper fit of the $\Delta R/R$ spectra. These correspond to four bandgaps that were labeled E_{a1}, E_{b1}, E_{a2}, E_{b2}. These multiple bandgaps arise because the entire thickness of the absorber is probed whereas in photoreflectance, one is limited to a thin top layer. For the CIGS layers with $x = 0.04$ and 0.12, the two stage deposition process resulted in a graded composition of Ga in the absorber. There are effectively two sub-layers with one richer in Ga and another poorer in Ga. Since each has basically the same CIGS band structure, there are two sets of bandgaps for each sub-layer, E_{a1}, E_{b1} and E_{a2} and E_{b2}.

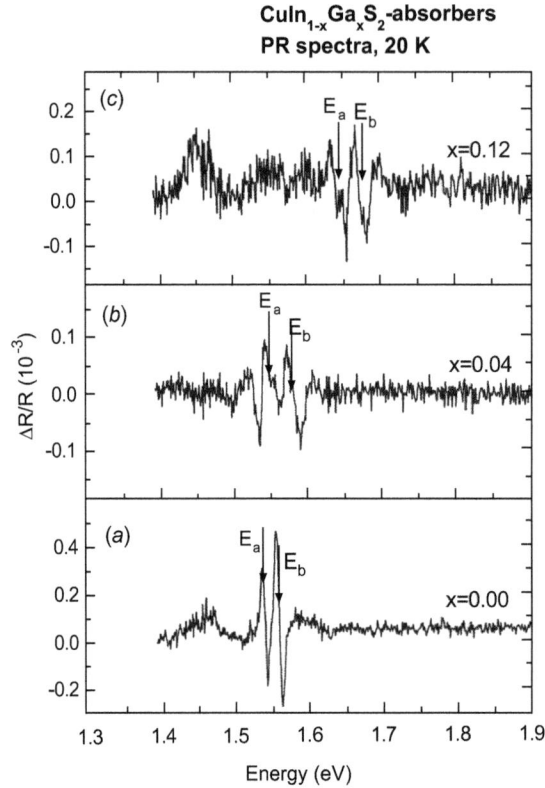

Figure 6: Low temperature (20K) photoreflectance spectra of $CuIn_xGa_{1-x}S_2$ absorber layers. In (a) and (b) the modulation source is a 488 nm Ar^+ laser and in (c) the modulation source is a 457.9 nm Ar^+ laser [10]. Reprinted with permission from 'Band-gap energies and strain effects in $CuIn_{1-x}Ga_xS_2$ based solar cells", S Theodoropoulou, D Papadimitriou, A G Mamalis, D E Manolakos, R Klenk and M-Ch Lux-Steiner, Semicond. Sci. Technol. 22, 933-940, August 2007, IOP Publishing. DOI: 10.1088/0268-1242/22/8/019.

The different Ga fraction inside the two sub-layers implies different lattice constants and strain in the absorber. The amount of lattice mismatch strain can be deduced using bandgaps measured by electroreflectance. Briefly, the change in the bandgaps ΔE_a and ΔE_b with respect to the bulk (unstrained) GIGS with the same composition are first determined. Then by using the hydrostatic deformation potential, a and shear deformation potential, b and the elastic stiffness constants C_{11} and C_{12}, the strain to first order can be found from [10]:

$$\Delta E_a = \left[-2a \left(\frac{C_{11} - C_{12}}{C_{11}} \right) - b \left(\frac{C_{11} + 2C_{12}}{C_{11}} \right) \right] \varepsilon \qquad (4.16)$$

$$\Delta E_b = \left[-2a \left(\frac{C_{11} - C_{12}}{C_{11}} \right) + b \left(\frac{C_{11} + 2C_{12}}{C_{11}} \right) \right] \varepsilon$$

By using literature values for the deformation potentials and elastic stiffness constants, the strain within the CIGS absorber was calculated to be under 0.7% for x = 0.04 and under 0.2% for x = 0.12. The calculated strain depends on the literature data values and the bandgap used.

In addition to GIGS, cadmium telluride CdTe, a II-VI compound semiconductor is also an important thin film photovoltaic material. Alejo-Armenta and co-workers at the Universidad Automnoma de Sinaloa in Mexico used the photoreflectance technique to study the energy band structure of oxygenated CdTe thin films [11]. The films were sputter deposited from a CdTe target in an argon oxygen nitrogen plasma. The photoreflectance measurement system used consisted of a tungsten halogen lamp with monochromator and the pump beam is an attenuated Ar laser modulated at a frequency of 213 Hz. By fitting to the measured $\Delta R/R$ spectrum, the fundamental energy gap of the oxygenated CdTe could be determined. The measured band gap is slightly smaller than the 1.5 eV band gap for unstressed CdTe because of internal tensile stress [11]. Other stress related effects of the band structure were also studied.

5. SUMMARY

Recent developments in photoreflectance spectroscopy had resulted in much faster measurements of the photoreflectance spectra of semiconductor substrates. This has made possible in line characterization of strained silicon wafers. As with the spectroscopic ellipsometry technique, the method required prior knowledge of the deformation potentials and elastic properties of the semiconductor being measured. Due to the use of morphic effects, the application of this method is dependent on the availability of this material data.

REFERENCES

[1]   H. Chouaib, M.E. Murtagh, V. Guenebaut, S. Ward, P.V. Kelly, M. Kennard, Y.M. le Vailant, M.G. Somekhm M.C. Pitter and S.D. Sharples, "Rapid photoreflectance spectroscopy for strained silicon metrology," *Rev. Sci. Instrum.,* vol. 79, pp. 1031061-3, Oct. 2008.

[2]   J.I. Pankove, *Optical Processes in Semiconductors,* Dover: New York, 1971.

[3]   F.H. Pollak and H. Shen, "Modulation spectroscopy of semiconductors: bulk/thin film, microstructures, surfaces/interfaces and devices," *Mater. Sci. Eng. R*, vol. 10, pp. 275-374, Sep. 1993.

[4]   M. Geddo, G. Guizzetti, M. Patrini, T. Ciabattoni, L. Seravalli, P. Frigeri and S. Franchi, "Metamporphic buffers and optical measurements of residual strain," *Appl. Phys. Lett.,* vol. 87, pp. 263120-1-3, Dec. 2005.

[5]   D.E. Aspnes, "Third derivative modulation spectroscopy with low-field electroreflectance," Surf. Sci., vol. 37, pp. 418-442, Jun. 1973.

[6]   W. Chism, D. Pham and J. Allgair, "Photo-Reflectance characterization of nanometer scale active layers in Si," In: *Frontiers of characterization and metrology for nanoelectronics*, 2007, pp. 64-68.

[7]   Q. Li, H.H. Tan and C. Jagadish, 'A new optical front-end compensation technique for suppression of spurious signal in photoreflectance spectroscopy using an antiphase signal," *Rev. Sci. Instrum.,* vol. 81, pp. 0431021-4, Apr. 2010.

[8]   J. Munguia, H. Chouaib, J. de la Torre, G. Bremond, C. Bru-Chevallier, A. Sibai, B. Champagnon, M. Moreau and J.-M. Bluet, "Optical strain measurement in ultrathin sSOI wafer," *Nucl. Instru. Meth. Phys. Res. B*, vol. 253, pp. 18-21, Dec. 2006.

[9]   J. Munguia, J.-M. Bluet, H. Chouain, G. Bremond, M. Mermoux and C. Bru-Chevallier, "Strain dependence of the direct bandgap in thin silicon on insulator layers," *J. Phys. D: Appl. Phys.*, vol. 43, pp. 255401, Jun. 2010.

[10]   S. Theodoropoulou, D. Papadimitrioum, A.G. Mamalis, D.E. Manolakos, R. Klenk and M-Ch Lux-Steiner, "Band-gap energies and strain effects in $CuIn_{1-x}Ga_xS_2$ based solar cells," *Semicond. Sci. Technol.* vol. 22, pp. 933-940, Aug. 2007.

[11]   L.N. Alejo-Armenta, F.J. Espinoza-Beltran, C.A. Alejo-Armenta, C. Vazquez-Lopez, H. Arizpe-Chavez, R. Ramirez-B0n, O. Zelaya-Angel and J. Gonzalez-Hernandez, "Strain effects on the energy band-gap in oxygenated CdTe thin films studied by photoreflectance," *J. Phys. Chem. Solids,* vol. 60, pp. 807-811, Jun. 1999.

CHAPTER 5

Micro-Raman Spectroscopy

Abstract: The use of inelastic scattering of laser light or Raman scattering to measure strain with micron scale spatial resolution is reviewed in this chapter. The principle of Raman scattering is discussed first and is followed by the effect of strain on the frequency shift of Raman scattered light. The experimental setup for performing Raman scattering measurements and the needed experimental precautions are given in detail. The method is illustrated by examples from strained silicon, polycrystalline silicon and strain fields near device isolation structures in crystalline silicon.

Keywords: Phonons, Inelastic scattering, Raman shift, Wavenumber, Polysilicon.

1. INTRODUCTION AND BACKGROUND

This chapter describes a widely used non-destructive local strain measurement technique that is known as micro-Raman spectroscopy. This technique is closely related to the more recent tip-enhanced Raman spectroscopy that will be discussed in chapter 9. All forms of Raman spectroscopy is fundamentally an inelastic optical scattering technique [1]. By measuring the frequency of scattered photons relative to the frequency of the incident monochromatic photons (from a laser), one can deduce useful information about the sample because the frequency shift of the inelastically scattered photons is dependent on the properties of the sample such as chemical bonding, temperature and stress.

The Raman effect was discovered by the Indian physicist and Nobel laureate Chandrasekhara Venkata Raman of the University of Calcutta in 1928 [1]. Raman observed this effect with his research collaborator K.S. Krishnan in the course of their study of optical scattering in liquids. As in other discoveries in science, Raman and Krishnan were not the only physicists who observed inelastic light scattering in condensed matter. At around the same time, two Russian academics G.S. Landsberg and L.I. Mandelstam of Moscow State University independently observed the inelastic optical scattering and the new phenomenon was called in the then Soviet Union to be combinatorial scattering of light. However, today the standard terminology is the Raman effect. C.V. Raman was awarded the Nobel prize for physics in 1930 for this discovery.

Raman spectroscopy was initially applied in analytical chemistry as a complementary vibration spectroscopy to Fourier transform infra-red (FTIR) spectroscopy [2]. By measuring the Raman spectrum of a material sample and comparing its Raman shift to the known shifts of tabulated chemical bonds, it is possible to deduce the types of chemical bonds present in a compound or in thin films. In nanoscience and nanotechnology, the Raman spectroscopy technique is especially useful to the study of the covalent bonding in the various allotropes of carbon such as diamond, fullerenes, carbon nanotubes and graphene [3].

Chemical bonding, however, is not the only factor that influences the Raman spectrum of a sample. In 1970, Anastassakis and co-workers at Brown University found that mechanical stress can also affect the amount of Raman shift of a material [4]. Thus if the chemical composition of the sample is known, one can use the Raman spectrum to quantitatively deduce the amount of stress and then apply Hooke's law to calculate the strain. In the next section, we first show the quantitative classical theory behind Raman spectroscopy and then proceed to explain the effect of stress on Raman scattering.

2. MEASUREMENT PRINCIPLE

2.1. Origin of Raman Scattering

It is useful to first understand the origin of Raman scattering in a crystal. The rigorous approach would be to use quantum mechanics. However, since this involves the use of Feynman diagrams [5], a simpler classical approach which is nonetheless sufficient for the present purpose is given instead [5]. This theory is applicable to both micro-Raman and tip enhanced Raman scattering. The Raman scattering effect arises because at non-zero temperatures, the atoms inside any crystal are vibrating about some equilibrium position. This can be taken as the position of the atom at absolute zero temperature. The atomic vibrations

Terence K.S. Wong

are collective and each such vibration is called a normal mode. Furthermore, the energy of the normal modes is restricted to integer multiples of a quantum of energy called the phonon [6]. Mathematically, the collective vibrations can be written down using a normal coordinate $Q_j(r)$ for the jth phonon mode at position r:

$$Q_j(r) = A_j \exp[\pm i(q_j \bullet r - \omega_j t)] \tag{5.1}$$

Here, A_j is the amplitude; q_j is the wavevector and ω_j is the angular frequency of the jth normal mode. As a result of this collective vibration, the electric susceptibility of the crystal will also change periodically. The electric susceptibility describes how readily an incident electromagnetic field can induce an electric dipole moment in the crystal and is given the symbol χ. Since the normal coordinates are assumed to be small, the electric susceptibility can be expanded as a Taylor's series about the lattice equilibrium position with respect to Q_j as:

$$\chi = \chi_0 + \left(\frac{\partial \chi}{\partial Q_j}\right)_0 Q_j + \left(\frac{\partial^2 \chi}{\partial Q_j \partial Q_k}\right)_0 Q_j Q_k + \ldots \tag{5.2}$$

Note that in the above equation, there will be more than one first order term on the right hand side if the crystal has more than one normal mode. Similarly, there will be multiple second order terms. The induced electric dipole moment P is given by:

$$P = \varepsilon_0 \chi E \tag{5.3}$$

If one considers a monochromatic beam with frequency ω_i and wavevector k_i, the induced dipole moment will be:

$$P = \varepsilon_0 \chi \bullet E_0 \exp\left[i(k_i \bullet r - \omega_i t)\right] \tag{5.4}$$

If only the zero and first order terms of equation (5.2) are substituted into equation (5.4), then this can be rewritten as:

$$P = \varepsilon_0 \chi_0 \bullet E_0 \exp\left[i(k_i \bullet r - \omega_i t)\right] + \varepsilon_0 E_0 \left(\frac{\partial \chi}{\partial Q_j}\right)_0 A_j \exp\left[-i(\omega_i \pm \omega_j)t\right] \exp\left[i(k_i \pm q_j) \bullet r\right] \tag{5.5}$$

This equation states that there are three main components of the time dependent induced dipole moment. Since oscillating dipoles will re-radiate light in all directions, these dipole moments will give rise to three components of scattered light. The first one is Rayleigh scattering for which there is no change in the frequency of the light. This is the elastic scattering component that gives rise to the blue color of the sky during daytime. The other two components are the Raman scattered light. These have a slightly shifted frequency with respect to the incident light and the amount of shift is equal to the characteristic frequency of the normal mode. The components with frequencies, $\omega_i + \omega_j$ and $\omega_i - \omega_j$ are called the anti-Stokes and Stokes Raman scattering respectively.

An important concept to appreciate is that for a given incident electromagnetic wave, not all the available normal modes of the crystal may give rise to observable Raman scattered light simultaneously. This is due to the so-called polarization selection rules [5]. The Raman scattering efficiency, I can be defined for the polarization vectors of an incident light wave, e_i and a scattered wave, e_s as:

$$I = C \sum_j \left| e_i \bullet R_j \bullet e_s \right|^2 \tag{5.6}$$

Here, C is a constant and R_j are the Raman tensors of the crystal. The Raman tensors are of rank 2 and they are proportional to the electric susceptibility. For a cubic system such as silicon, the three Raman tensors are given by [7]:

$$R_x = \begin{bmatrix} 0 & 0 & 0 \\ 0 & 0 & d \\ 0 & d & 0 \end{bmatrix} \quad R_y = \begin{bmatrix} 0 & 0 & d \\ 0 & 0 & 0 \\ d & 0 & 0 \end{bmatrix} \quad R_z = \begin{bmatrix} 0 & d & 0 \\ d & 0 & 0 \\ 0 & 0 & 0 \end{bmatrix} \tag{5.7}$$

The coordinate system chosen is that $x = [100]$; $y = [010]$ and $z = [001]$. The non-zero component d within R_j is based on the notation used by Loudon and is one component of the polarizability tensor [13]. By substituting these tensors into equation (5.6), it will be seen that for certain geometries, the scattering efficiency is zero. This results in the selection rules of Raman scattering and they will be mentioned again in chapter 9. The selection rules are important because they restrict the choice of Raman modes for strain characterization.

2.2. Effect of Mechanical Stress on Raman Modes

When stress is applied to a semiconductor crystal, the change in the lattice parameters can lead to a change in the characteristic frequencies of the phonon modes. If the symmetry of the crystal lattice is reduced, there can be a lifting of the degeneracies in the phonon modes. These changes often result in a shift in the peak of the Raman spectrum and can be used to infer the level of stress. The quantitative theory for the effect of stress on Raman spectra was first studied by Ganesan *et al.* in 1970 [5]. The main result is that the shift in the three Raman mode frequencies can be found by solving the secular equation:

$$\begin{vmatrix} p\varepsilon_{11} + q(\varepsilon_{22} + \varepsilon_{33}) - \lambda & 2r\varepsilon_{12} & 2r\varepsilon_{13} \\ 2r\varepsilon_{12} & p\varepsilon_{22} + q(\varepsilon_{33} + \varepsilon_{11}) - \lambda & 2r\varepsilon_{23} \\ 2r\varepsilon_{13} & 2r\varepsilon_{23} & p\varepsilon_{33} + q(\varepsilon_{11} + \varepsilon_{22}) - \lambda \end{vmatrix} = 0 \tag{5.8}$$

The coefficients p, q and r in the secular equation are called phonon deformation potentials and ε_{ij} are the components of the strain tensor. λ represents the eigenvalues of the secular equation. Equation needs to be solved for the eignevalues and these are then used to calculate the difference in frequency of each Raman mode under stress and without stress.

$$\Delta\omega_j = \omega_j - \omega_{j0} \approx \frac{\lambda_j}{2\omega_{j0}} \tag{5.9}$$

We consider two cases relevant to the semiconductor industry as examples. The first one is uniaxial stress σ applied along the [100] direction. The shifts in the frequencies of the three Raman modes of silicon are found to be [5]:

$$\Delta\omega_1 = \frac{\lambda_1}{2\omega_0} = \frac{(pS_{11} + 2qS_{12})\sigma}{2\omega_0}$$

$$\Delta\omega_2 = \frac{\lambda_2}{2\omega_0} = \frac{(pS_{12} + q(S_{12} + S_{11}))\sigma}{2\omega_0} \tag{5.10}$$

$$\Delta\omega_3 = \frac{\lambda_3}{2\omega_0} = \frac{(pS_{12} + q(S_{12} + S_{11}))\sigma}{2\omega_0}$$

The angular frequency ω_0 is the characteristic frequency for the three degenerate modes of silicon and they each have the value of $\omega_0 = 520$ Rcm^{-1} (relative cm^{-1}, see below for definition). S_{11}, S_{12} are the components of the elastic compliance tensor of silicon and have the values [5]: $S_{11} = 7.68$ x 10^{-2} Pa^{-1} and $S_{12} = -2.14$ x 10^{-12} Pa^{-1}. The phonon deformation potentials have the values: $p = -1.43\omega_0^2$ and $q = -1.89\ \omega_0^2$. Depending on the geometric of the incident light, one can use equation (5.10) and the parameter values above to calculate the uniaxial stress from the shift in the characteristic Raman frequency. Thus, for the case of a backscattering geometry from a (001) silicon surface, the polarization selection rule shows that only the third Raman mode in equation (5.10) is observable. After substitution of the silicon parameter values, we obtain a simple relation for the Raman shift of (001) silicon [5]:

$$\Delta\omega_3\left(cm^{-1}\right) = -2\times10^{-9}\,\sigma\left(Pa\right) \tag{5.11}$$

Once the uniaxial stress is found, Hooke's law and the Poisson's ratio can be applied to find the three normal strain components ε_{xx}, ε_{yy} and ε_{zz}. The second case is that of a biaxial stress in the x-y plane. Considering the same backscattering geometry from a (001) surface, the third relation in (5.10) becomes [5]:

$$\Delta\omega_3\left(cm^{-1}\right) = -4\times10^{-9}\left(\frac{\sigma_{xx}+\sigma_{yy}}{2}\right)\left(Pa\right) \tag{5.12}$$

3. EXPERIMENTAL SETUP FOR MICRO-RAMAN SPECTROSCOPY

Fig. **1** shows the schematic diagram of the experimental setup for micro-Raman spectroscopy [5, 7]. Although Raman spectroscopy can be performed without using an optical microscope, it is often the case for semiconductor characterization that a focused light beam is used to acquire the Raman spectrum. Usually a monochromatic laser light source is used for the incident beam. In the early days of micro-Raman spectroscopy, visible light from ion lasers are often used. However, as the need for surface selectivity increases with the use of ultrathin layers such as strained silicon layers on a substrate, the use of ultraviolet lasers has become more and more common in recent years. The laser light is coupled to a confocal optical microscope *via* an optical fiber and the light is focused by an objective lens. In confocal microscopes, the beam spot can be varied from about 1μm to several μm. Due to microscope design, the light beam is usually incident normally onto the sample from above. For this geometry, the scattered light is usually collected in the backscattered mode using the same objective lens. However, for transparent samples and for certain samples, the light beam can be incident from below for what are called inverted optical microscopes.

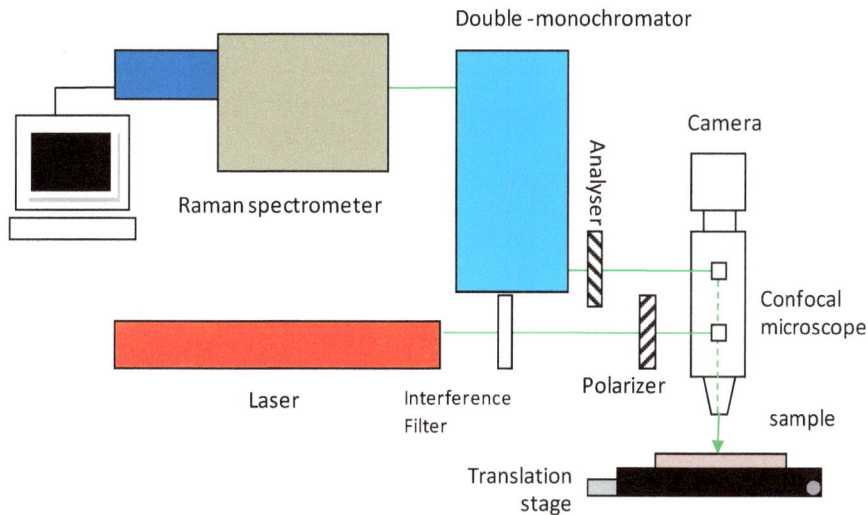

Figure 1: Schematic of the experimental setup for a typical micro-Raman spectroscopy system.

The incident light is scattered by the sample by the Rayleigh and Raman mechanisms. Since the inelastically scattered photons are only slightly shifted in frequency with respect to the incident and the Rayleigh scattered light and the Raman effect is intrinsically weak, it is essential to have a highly selective spectrometer to analyze the scattered light and to generate usable Raman spectra. A typical arrangement is to use a double monochromator consisting of two identical monochromators in series. The two monochromators are separated by an intermediate slit. There is also an entrance and an exit slit to the double monochromator with variable slit widths. Each monochromator consists of a diffraction grating and two mirrors. The function of the monochromator is to select a wavelength of interest and to reject light from neighboring wavelengths and stray light from entering the detector device. The use of two or more monochromator stages ensure that stray light is reduced to a minimum. The selected wavelength is then directed to the spectrometer for spectral collection. The spectrometer has a spectrograph and a light detector. The spectrograph is basically similar to the monochromator and is used in this case to disperse the incident light into different wavelengths which are then spatially separated. The intensity of the light can be detected by multichannel detectors, charge coupled devices or photomultiplier tubes.

An important advantage of Raman spectroscopy is that the method does not require sample preparation and is non-invasive. A user can simply put the sample onto the sample stage and use the translation devices to move the laser spot to a point of interest on the sample. A spectrum can be collected automatically with the hardware under computer control. Although the experimental procedure is relatively straightforward, one nevertheless needs to be careful to obtain usable Raman spectra. An important consideration is unreliable Raman shifts due to drifts in the laser wavelength or the spectrometer itself. All Raman spectra collected are basically plots of intensity *versus* the Raman shift. This is a change in photon frequency relative to the incident laser frequency and is expressed in the unit of relative wavenumber Rcm^{-1} (relative cm^{-1}). In the literature, the unit is also often written as cm^{-1} only. If the laser wavelength varies during a measurement, the measured frequency shift will be in error. The incident laser power is another important consideration. If the incident power is high, the sample could be heated up and a Raman down-shift due to purely temperature effects will occur. In practice, it is necessary to take necessary precautions or make corrections to the acquired spectrum prior to strain analysis.

4. APPLICATION EXAMPLES

In this section, we discuss examples on the application of micro-Raman spectroscopy for strain characterization in semiconductor thin film materials. Since this technique had been applied since the 1990's, there exists a rather wide body of literature and as a result, only selected papers will be reviewed. The more recent work on strained silicon and silicon based materials will be described first because these are of more contemporary interest. This will be followed by a review of earlier work on strain effects in bulk silicon due to dielectric isolation structures. This work is also important because it is what motivates semiconductor technologists to develop the micro-Raman technique.

4.1. Biaxial Strained Silicon

Fig. **2** shows the Raman spectra for three strained silicon samples S15, S20 and S25 [8]. These were the samples that were discussed in chapter 3. The shift of the Si-Si transverse optical phonon peak in the strained silicon relative to bulk silicon measured separately is related to the in-plane strain *via* (5.13):

$$\Delta\omega_{Si-Si}^{\varepsilon-Si} = \omega_{Si-Si}^{bulk} - \omega_{Si-Si}^{\varepsilon-Si} = b_{Si-Si}^{\varepsilon-Si}.\varepsilon_{xx}^{\varepsilon-Si} \tag{5.13}$$

where the strain-shift coefficient, $b_{Si-Si}^{\varepsilon-Si}$ has the value -784 cm^{-1} [8]. By using equation (5.13), the strain ε_{xx} are found to be 0.51%, 0.77% and 1.03% for samples S13, S20 and S25 respectively. These are close to values found by spectroscopic ellipsometry (chapter 3) and X-ray diffraction (chapter 11).

Perova and co-workers at the University of Dublin used the micro-Raman technique in 2006 to characterize both strain and stress in biaxial strained silicon grown on ultra thin silicon germanium virtual substrates [9]. The metamorphic substrates used for growing biaxial strained silicon as described in chapter 1 are typically quite thick with a total thickness of several microns. Apart from the low process throughput due to this, there is the problem of heat dissipation in the strained silicon caused by the lower thermal conductivity of silicon germanium relative to silicon. It should be noted that silicon germanium is used as a thermoelectric material.

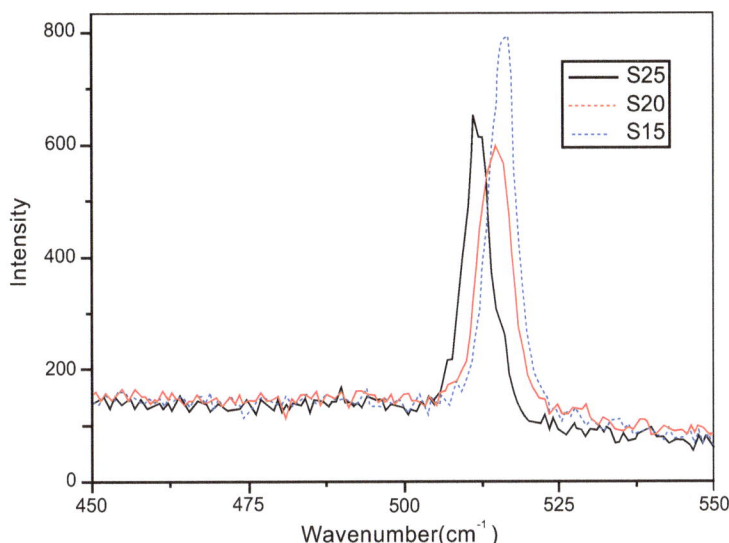

Figure 2: Raman spectra of samples S15, S20 and S25 for source wavelength of 325nm. Reprinted with permission 'Characterization of biaxial stressed silicon by spectroscopic ellipsometry and synchrotron X-ray scattering', T.K.S. Wong, Y. Gong, P. Yang, C.M. Ng, Semicond. Sci, and Technol. 22, 1232-1239, November 2007, IOP Publishing DOI: 10.1088/0268-1242/22/11/009.

In order to circumvent the heat transfer problem, an ultra-thin silicon germanium virtual substrate was used for growing the strained silicon. The total thickness of the Si_xGe_{1-x} buffer layer is under 80nm. This layer is divided into two sub-layers. The first Si_xGe_{1-x} sub-layer is 20-30nm thick and is grown by molecular beam epitaxy on (100) silicon at very low temperature [9]. The germanium content in the buffer layer ranged from 12 at.% to 42 at.%. Above the Si_xGe_{1-x} buffer layer is a Si_xGe_{1-x} intermediate layer with the same composition and a thickness of 120nm. Thus, the total Si_xGe_{1-x} layer thickness is much thinner than the conventional process. The strained silicon is grown at 500°C with a thickness of 10 – 20nm. A series of such layers with different concentration of germanium in the ultra-thin virtual substrate was grown.

Strain was characterized by micro-Raman spectroscopy using a Renishaw 1000 system with two excitation wavelengths (514nm and 325nm). Since visible wavelength has a greater penetration depth, the entire film stack including the (100) silicon substrate can be probed for these ultra thin silicon germanium virtual substrates. This is different from strained silicon grown on a conventional virtual substrate that is much thicker.

When excited by visible light, the Raman spectrum of strained silicon grown on ultra-thin silicon germanium virtual substrates consists of three Si-Si peaks in the frequency shifted region of 490 cm^{-1} – 530 cm^{-1} (Fig. **3a**). These peaks can be de-convoluted by fitting three Lorentzian functions (Fig. **3b**). The most prominent peak is that of the unstrained silicon substrate at 520 cm^{-1}. The side peak in the region of 510 cm^{-1} – 518 cm^{-1} is due to the top strained silicon layer and there is a third peak in the region of 490 cm^{-1} – 510 cm^{-1} which is due to the Si-Si Raman mode in the Si_xGe_{1-x} layer. The intensity of this third peak decreases with increasing germanium concentration. The Raman peak of the unstrained substrate is useful because the position of this peak is constant and can be used as an internal marker for the measurement of relative peak shift in the strained silicon. Four samples with germanium concentration of $x = 0.12, 0.21, 0.32$ and 0.42 in the buffer layer were thus measured and the Raman shifts (514nm excitation) were deduced by peak fitting. These same samples were measured using 325nm ultraviolet excitation. The penetration depth is much reduced and only the peak due to the strained silicon is detected. By using the plasma lines of the laser as reference, the Raman shift for the strained silicon was deduced. These were in agreement with the shifts obtained by 514nm excitation.

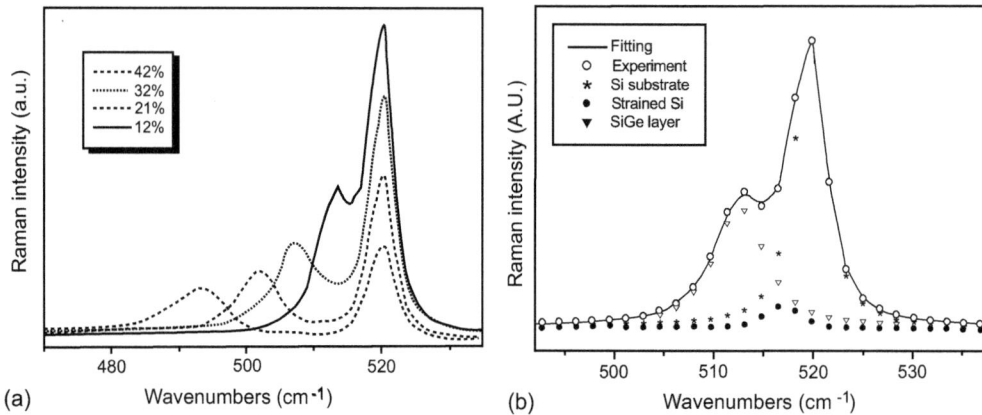

Figure 3: (a) Visible micro-Raman spectra of strained silicon on ultra-thin SiGe buffer with different Ge content; (b) Fitting of one Raman spectrum by three Lorentzian functions [9]. Reprinted from *Mater. Sci. Eng. B,* vol. 135, T.S. Perova, K. Lyutovich, E. Kasper, A. Waldron, M. Oehma and R.A. Moore, "Stress determination in strained-Si grown on ultra-thin SiGe virtual substrates", pp. 192-194, Copyright 2006, with permission from Elsevier.

The strain within the strained silicon is found from the measured Raman peak shift, $\Delta\omega$ by applying:

$$\varepsilon_{Si}\left(\%\right) = \left(\frac{\Delta\omega}{b}\right) \times 100\% \qquad\qquad (5.14)$$

In this equation, b is the strain-phonon coefficient and has a value of 930 cm^{-1} [9]. The stress of the strained silicon is calculated from the Young's modulus and Poisson's ratio.

Himcinschi and colleagues at the Max Planck Institute of Microstructure Physics used a similar method to study the strain in strained silicon nanostructures patterned by electron beam lithography [10]. A strained silicon layer was etched into arrays of 150 nm x 150 nm and 150 nm x 750 nm pillars. A UV micro-Raman spectroscopy system was used to measure the Raman spectra of these nanostructures because of the much smaller penetration depth. Since the beam spot is much greater than the size of each nanostructure, the measured Raman spectrum consisted of the Si-Si peak from strained silicon and the Si-Si peak from silicon germanium layer beneath. This means that in this case, the deduced strain is averaged over all the strained silicon nanopillars that are probed by the beam. A strain measurement on an individual pillar is impossible with this instrument. The reported strain in the 150 nm x 150 nm nanopillar is 0.45% and that in the 150 nm x 750 nm pillars is 0.63% [10]. These strain values are smaller than the strain of the initial layer of about 0.95%. Thus, the smaller the nanostructure, the greater is the degree of strain relaxation.

4.2. Polycrystalline Silicon

Until 2007, polycrystalline silicon (or polysilicon) was the standard gate electrode material for CMOS devices. In addition to CMOS applications, polysilicon is also an important material for MEMS systems. In 2004, Teixeira and co-workers at Unicamp in Brazil applied this technique to characterize stress in polycrystalline silicon films with the aim of developing low stress layers for MEMS applications [11]. Polycrystalline silicon deposited by conventional horizontal low pressure chemical vapor deposition (LPCVD) processes typically has high compressive stress. The stress in polysilicon can be due to the deposition process, defects in the film and film microstructure. Excessive stress can lead to bending or buckling of free standing polysilicon microstructures or cracking and delamination of films on substrates.

In order to minimize the stress in the polycrystalline silicon, a vertical LPCVD reactor was used for deposition. The gas flow into the vertical LPCVD reactor consisted of high purity hydrogen carrier gas and silane at a flow ratio of 120:1. The deposition occurred by thermal decomposition of silane. Low stress polysilicon was found to be obtained for deposition at 5 Torr and 750°C – 900°C and at 10 Torr and 700°C – 900°C. The polysilicon was deposited onto (100) silicon wafers with a thin layer of thermal oxide.

A Jobin-Yvon T64000 Raman spectrometer (514nm, 6mW) was used to characterize the stress and the crystallinity of the polysilicon. Raman spectra were collected in the backscattered geometry. Since the polysilicon films deposited have low stress, the location of the Raman peak is near to the 520 cm^{-1} peak for single crystal silicon. For stress determination, the Raman spectra were de-convoluted by fitting two Lorentzian functions to account for the amorphous and crystalline constituents of the film. The peak of the Lorentzian for the crystalline component relative to the peak at 520 cm^{-1} for single crystal silicon was used to calculate the stress in the polysilicon. The equation used is:

$$\sigma = -250(MPa/cm) \times \Delta\omega \tag{5.15}$$

Here, σ is measured in MPa and $\Delta\omega$ is the shift of the fitted crystalline silicon peak in polysilicon relative to that of single crystal silicon. By using the above equation, the stress in all the deposited polysilicon was found to be tensile and it decreased rapidly with increasing deposition temperature up to 800°C and then became stable at a minimum value. The stress behavior in this series of polysilicon films was reported to be very different from conventional LPCVD processes. Usually, a compressive stress was observed in polysilicon films and in order to reduce this stress, the films have to be furnace annealed after deposition.

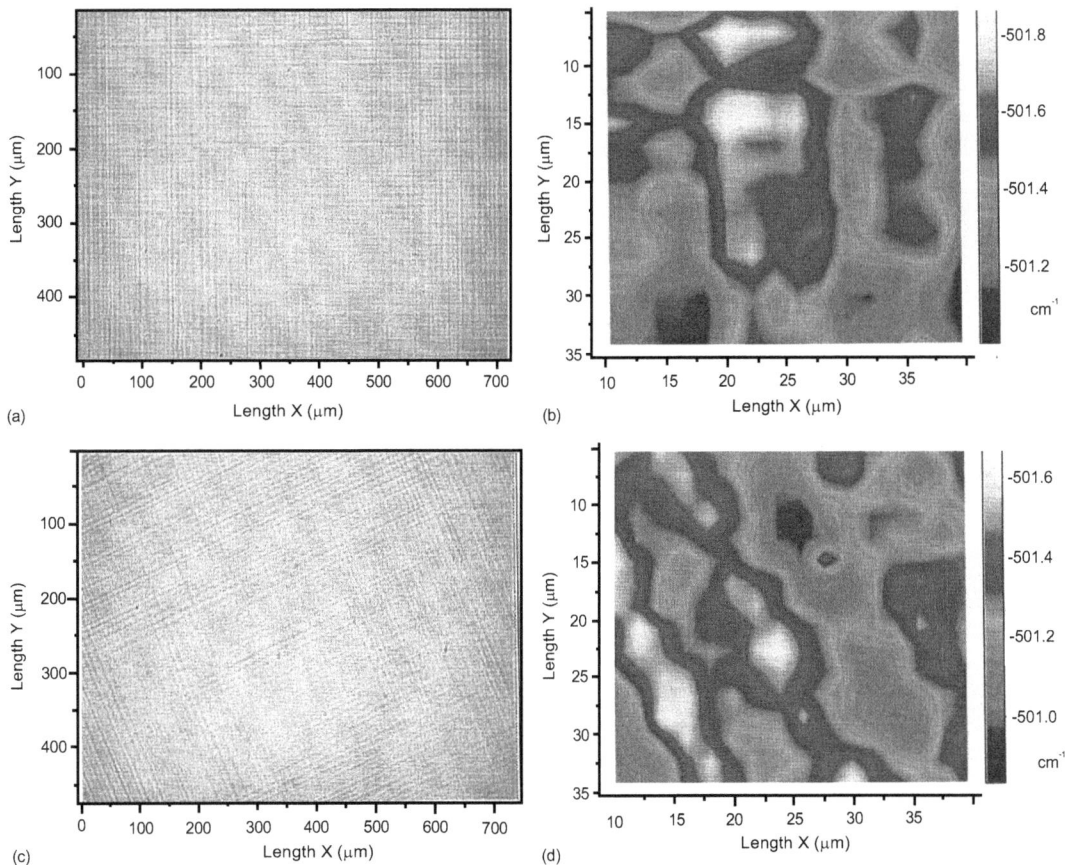

Figure 4: (a) Optical image of strained silicon on Si$_{1-x}$Ge$_x$; (b) Raman map of Si-Si band of an area within the optical image in (a); (c) optical image of another area of the sample after rotation by 45° (d) Raman map of Si-Si band of an area within the optical image in (c) [12]. Reprinted from *Mater. Sci. Semicond. Proc.* vol. 8, G.G. Goodman, V. Pajcini, S.P. Smith and P.B. Merrill, "Characterization of strained Si structures using SIMS and visible Raman", pp. 255-260, Copyright 2005, with permission from Elsevier.

4.3. Silicon Germanium

Micro-Raman spectroscopy had also been used to measure strain variations across the nominally relaxed Si$_{1-x}$Ge$_x$ substrates used for growing strained silicon. It is important to carry out such measurements because any residual strain variations in the $Si_{1-x}Ge_x$ can lead to strain (and mobility) variations in the strained silicon layer. Goodman and co-workers at the Evans analytical group used both optical microscopy and visible Raman spectroscopy (514.5nm, 1mW) to obtain the results shown in Fig. **4** [12]. The optical image of Fig. **4a** shows the characteristics cross hatch pattern seen on the surface of a strained silicon layer due to misfit dislocations in the underlying $Si_{1-x}Ge_x$. The Raman map of Fig. **4b** shows the Si-Si Raman band as a function of position within a 30µm x 30µm area of the image in Fig. **4a**. The contrast in the Raman map implies residual stress in the $Si_{1-x}Ge_x$ and a variation of +/-0.05% strain in the $Si_{1-x}Ge_x$ was deduced [12]. When the sample is rotated by 45°, the Raman map follows the optical image. The lateral width of the stress variations in Fig. **4c** and Fig. **4d** corresponds to the cross hatch pattern width in the optical images. This suggests that assuming the composition of Ge is uniform, the strain variation is associated with the dislocations in the $Si_{1-x}Ge_x$ layer.

4.4. Isolation Structures

Isolation structures are used to electrically separate semiconductor devices (MOS and bipolar) inside integrated circuits. Initially, local oxidation of silicon (LOCOS) and its variants were used. However, today trench isolation structures are used because they are more suited for aggressive scaling. In trench isolation, a silicon trench is first etched in the substrate and is refilled by SiO_2. After refilling, stress is often induced in the silicon adjacent to a trench and this can result in lattice defects and leakage current. As a result, micro-Raman spectroscopy is used to study the stress in silicon due to trench isolation. In one study [5], the trench isolation structure is cleaved and micro-Raman measurements are made at different depths of the cleaved surface between two trenches. The stress was calculated from the Raman shift by assuming uniaxial stress. Itoh *et al.* [5] had carried out a similar study for three different dielectric deposition processes and compared the stresses induced by each. The study of stress in isolation structures by micro-Raman spectroscopy is historically important because it led on to the idea of strained silicon channels in MOS devices.

REFERENCES

[1]   M.S. Amer, Raman Spectroscopy, Fullerenes and Nanotechnology, RSC publishing: Cambridge, 2010.

[2]   R.D. Braun, Introduction to Instrumental Analysis, Mc-Graw Hill: New York, 1987.

[3]   G. Picardi, M. Chaigneau and R. Ossikovski, "High resolution probing of multi wall carbon nanotubes by tip enhanced Raman spectroscopy in gap-mode," *Chem. Phys. Lett.,* vol. 469, pp. 161-165, Feb. 2009.

[4]   E. Anastassakis, A. Pinczuk, E. Burstein, F.H. Pollak and M. Cardona, "Effect of static uniaxial stress on the Raman spectrum of silicon," *Solid State Comm.,* vol. 8, pp. 133-138, Jan. 1970.

[5]   I. De Wolf, "Micro-Raman spectroscopy to study local mechanical stress in silicon integrated circuits," *Semicond. Sci. Technol.,* vol. 11, pp. 139-154, Feb. 1996.

[6]   C. Kittel, Introduction to Solid State Physics, Wiley: New York, 1996.

[7]   I. De Wolf, H.E. Maes and S.K. Jones, "Stress measurements in silicon devices through Raman spectroscopy: Bridging the gap between theory and experiment," *J. Appl. Phys.,* vol. 79, pp. 7148-7156, May. 1996.

[8]   T.K.S. Wong, Y. Gong, P. Yang and C.M. Ng, "Characterization of biaxial stressed silicon by spectroscopic ellipsometry and synchrotron x-ray scattering," *Semicond. Sci. Technol.,* vol. 22, pp. 1232-1239, Nov. 2007.

[9]   T.S. Perova, K. Lyutovich, E. Kasper, A. Waldron, M. Oehma and R.A. Moore, "Stress determination in strained-Si grown on ultra-thin SiGe virtual substrates," *Mater. Sci. Eng. B,* vol. 135, pp. 192-194, Dec. 2006.

[10]   C. Himcinschi, I. Radu, R. Singh, W. Erfurth, A.P. Milenin, M. Reiche, S.H. Christiansen and U. Gosele, "Relaxation of strain in patterned strained silicon investigated by UV Raman spectroscopy," *Mater. Sci. Eng. B,* vol. 135, pp. 184-187, Dec. 2006.

[11]   R.C. Teixera, I. Doi, M.B.P. Zakia, J.A. Diniz and J.W. Swart, "Micro-Raman stress characterization of polycrystalline silicon films grown at high temperature," *Mater. Sci. Eng. B,* vol. 112, pp160-164, Sep. 2004.

[12]   G.G. Goodman, V. Pajcini, S.P. Smith and P.B. Merrill, "Characterization of strained Si structures using SIMS and visible Raman," *Mater. Sci. Semicond. Proc.,* vol. 8, pp. 255-260, Feb. 2005.

[13]   R. Loudon, "The Raman effect in crystals," *Adv. Phys.,* vol. 13, pp. 423-482, Oct. 1964.

PART 3: ELECTRON BEAM STRAIN METROLOGY

In this part, we describe three methods of strain metrology that are based upon the use of electron microscopes. Either a finely focused electron beam in a scanning electron microscope (SEM) or a collimated electron beam in a transmission electron microscope (TEM) can be used to obtain strain information from a semiconductor sample. Unlike the optical methods in part 2, all three methods are suitable for probing device structures because electron beam techniques have much higher spatial resolution and the researcher can select the area of interest within the sample. The cathodoluminescence method of chapter 6 is based on the effect of strain upon the bandgap of a semiconductor. From the standpoint of principle, it is similar to the methods in chapters 3 to 5.

The other two methods in this part make use of the wave property of the electron. The electron diffraction methods of chapter 7 discuss two recent techniques: nanodiffraction and convergent beam electron diffraction. Both techniques had been applied to probe the strain distribution within strained silicon transistors. The electron holographic Moire method of chapter 8 is a very new and promising method. It makes use of the interference of two coherent electron beams from the same electron source in a TEM to generate an electron hologram. Unlike conventional holograms, one beam passes through a strained region of the sample and the other beam passes through an unstrained substrate region of the sample. Strain information is then extracted from the phase image derived from the electron hologram by the geometrical phase analysis method developed originally for high resolution TEM.

CHAPTER 6

Cathodoluminescence Method

Abstract: A strain characterization method with nanoscale spatial resolution called cathodoluminescence is explained. The method involves the use of a small electron probe with high energy. The energy loss by the primary electrons results in the emission of light which are collected by reflective optics and detected by a spectrometer. By measuring the cathodoluminescence spectrum during scanning, a spatial map of the stress distribution can be deduced. This is because stress affects the peak of the emission spectrum from the sample *via* the piezospectroscopic effect. By monitoring the peak shift, the strain can be deduced from the piezospectroscopic coefficients. This method is especially useful for ceramic materials such as oxides and glasses for which the Raman effect is extremely weak. However, stress distribution in epitaxial III-nitride semiconductors had also been studied.

Keywords: Cathodoluminescence, Scanning electron microscope, Piezospectroscopic effect, III-nitride, optical glasses.

1. INTRODUCTION

The first chapter of this section on electron beam methods had its origins in a related optical technique called optical microprobe piezospectroscopy (OMPS). This technique was first reported by Molis and Clarke at the IBM Thomas J. Watson Research Centre in 1990 [1]. Subsequently, it was extended to an electron beam probe for aluminum oxide by Ostertag *et al.* at the National Institute of Standards and Technology (Gaithursburg, Maryland) in 1991 [2]. Both the cathodoluminescence (CL) method and the OMPS method make use of the same physical principle to determine strain in a sample. The basic idea is to use an optical beam or an electron beam to excite optical emission or luminescence from a sample and use the effect of stress on the luminescence spectrum to deduce the strain. The relationship between the frequency shift of the peak of the luminescence spectrum and the applied stress is called the piezospectroscopic effect.

It is helpful to first briefly review luminescence in solid state materials. For the interested reader, there is an excellent text on this topic by Kitai [3]. Luminescence is one of several possible interactions between an energetic radiation or particle beam with a semiconductor sample (bulk or thin film). When a light beam with sufficient photon energy is incident on a semiconductor sample, its energy will be absorbed by the sample and this energy can be re-emitted as photons of characteristic energy. This comes about because of radiative electronic transitions from a higher energy level to a lower energy level. The same effect can be obtained by an electron beam. When an electron beam from a scanning electron microscope (SEM) penetrates the sample, it loses energy by multiple mechanisms. One of these is the emission of optical radiation and in this context is called CL. Other energy loss mechanisms include secondary electron emission and X-ray emission. Secondary electron emission is important for image formation and X-ray emission is important for chemical analysis of the sample. The main difference between CL and photoluminescence (PL) is that the diameter of the electron beam is generally much smaller and thus the CL signal is originated from a smaller generation volume. This difference gives CL method a higher spatial resolution.

There are two types of solid state luminescence. In intrinsic luminescence, the optical emission is due to radiative transitions between the energy levels of an ideal semiconductor. Examples of intrinsic luminescence include those from a direct band gap semiconductor such as GaN and AlGaAs. Certain conjugated (semiconducting) organic molecules are also strong intrinsic emitters and are nowadays used widely for flat panel displays. The luminescence phenomenon that is more relevant to the present discussion is extrinsic luminescence. Extrinsic luminescence arises from the impurities or crystal defects in a semiconductor. This phenomenon is in fact the operating principle of a type of robust display called thin film electroluminescent displays (TFEL) [3]. The active material in the TFEL is a phosphor layer that comprises a wide band gap semiconductor and impurity ions of a transition metal or a metal from the lanthanide series. The wide band gap semiconductor is merely a host for the impurity ions and does not possess intrinsic luminescence. The characteristics of the optical emission such as frequency and spectral shape are determined entirely by the impurity ion. As a result of this, the electronic transitions in extrinsic

Terence K.S. Wong

luminescence are intra-atomic and can take place within an electronic shell or between electronic shells of the impurity atom or ion. Such impurity ions are therefore also called luminescence centers. Although extrinsic luminescence is mainly used for display applications, it is also useful in strain metrology because of the piezospectroscopic effect.

Why should one care about the CL technique when there are already three other optical techniques (SE, PR and Raman) as described in part 2? The reason is that metal oxides are an increasingly important class of semiconducting materials with many device applications [4]. Metal oxides are ceramics and it is known that for ceramic materials, the Raman scattering effect is extremely weak [2]. Thus even with modern equipment, it may not be feasible to obtain useful Raman data from these materials. On the other hand, the metallic impurities in these materials readily give useful luminescence signals. Thus, when other optical techniques fail, one can consider the use of CL for high spatial resolution strain characterization.

2. CATHODOLUMINESCENCE IN SEMICONDUCTORS

When an energetic electron beam enters a semiconductor, the electrons will lose energy in the semiconductor through inelastic scattering events. In addition to secondary electrons and X-rays, the energy loss can result in the generation of electron hole pairs. The energy E_0 required to generate electron hole pairs is given by a semi-empirical relation proposed by Klein [5]:

$$E_0 = \frac{14}{5}E_g + E'$$

(6.1)

Here, E_g is the band gap of the semiconductor and E' is the energy lost to an integral number of phonons. The factor in front of E_g accounts for the energy band gap and the residual kinetic energy of the incident electron. This factor is determined from experiment by fitting to the CL data for a number of semiconductors [5]. The electron hole pairs are generated within a sphere of excitation with a radius called the range of excitation. This is not to be confused with the penetration depth which gives location of the centre of this sphere of excitation from the surface. The penetration depth d is a power law function of the energy of the electron beam and is approximately, $d \propto E^{\frac{3}{2}}$. These quantities are introduced because they can be used to estimate the number of electron hole pairs generated per unit volume N_d:

$$N_d = \frac{J}{1.6 \times 10^{-19}} \frac{E_p}{E_0} \frac{\tau}{d}$$

(6.2)

In equation (6.2), J is the current density of the electron beam; E_p is the energy of the primary electrons and τ is the electron hole pair lifetime. The ratio E_p/E_0 is the number of electron hole pairs created per incident electron. If the semiconductor is a direct band gap material, the recombination of these electrons and holes will lead to optical emission or CL. The external quantum efficiency η of this CL or the number of photons emitted from the semiconductor per incident electron is given by [5]:

$$\eta = \frac{(1-R)(1-\cos\theta_c)}{\left(1 + \frac{\tau_{rad}}{\tau_{non-rad}}\right)} \exp(-\alpha l)$$

(6.3)

In the previous equation, R is the reflectivity; θ_c is the critical angle for total internal reflection; α is the absorption coefficient of the semiconductor and l is the photon path length. $\tau_{rad}/\tau_{non-rad}$ is the ratio of the radiative minority lifetime to the non-radiative minority carrier lifetime.

3. PIEZOSPECTROSCOPIC EFFECT

The piezospectroscopic effect is the physical principle behind the CL and OMPS methods for strain characterization of ceramic materials. This effect was first observed by Ludwig Grabner in sintered alumina at the National Bureau of Standards, Washington D.C. in 1978 [6]. It refers to the effect of stress on the characteristic emission spectrum of impurity ions embedded in the material. This is most often manifested in the form of a frequency shift in the emission peak:

$$\Delta \upsilon = \pi_{ij}\sigma_{ij} \qquad\qquad (6.4)$$

Here, $\Delta \upsilon$ is the change in wavenumber of the luminescence peak and π_{ij} is the piezospectrospic coefficient. As with the stress tensor σ_{ij}, π_{ij} is a tensor of second rank. Since extrinsic luminescence originates from atomic energy levels, the piezospectroscopic effect can also be interpreted as the effect of stress on the electronic energy levels in the impurity atoms.

In the literature, the first and most widely studied material system for the piezospectroscopic effect is aluminum oxide (alumina) containing chromium (III) ions (Cr^{3+}) [2]. The Cr^{3+} ion functions as an efficient luminescence center in many light metal oxide materials such as alumina and magnesium oxide (MgO) because of its electronic configuration. The Cr^{3+} impurity ion occupies substitutional sites in the crystal lattice of the host oxide and is surrounded by an octahedron of oxygen ions. When Cr^{3+} is present in an oxide with a cubic crystal lattice such as MgO, it emits a characteristic spectrum with a single peak at 700nm. Since this is within the red portion of the visible spectrum, it is also called the ruby (R) line. The R line is due to the $^2E -> {}^4A_2$ electronic transition and the observed line shape of this line is best represented by the following generalized function [2]:

$$I\left(v-v_0\right) = G\left(v-v_0;W_1\right)\left[L\left(v-v_0;W_2\right)\right]^{\beta} \qquad\qquad (6.5)$$

Here G is a Gaussian function with width W_1 and L is a Lorentzian function with width W_2. υ_0 is the frequency of the peak and β is a positive exponent. This generalized function is used to mathematically fit for υ_0, W_1, W_2 and β.

In aluminum oxide, the Cr^{3+} ions are also surrounded by an octahedron of oxygen atoms. However, the crystal structure is not quite cubic and there is some strain of the octahedron along a trigonal direction. As a result, the upper ^2E level is split into two closely spaced energy levels and these give rise to two emission lines at 14,402.5cm^{-1} and 14,432.1cm^{-1} called the R_1 and R_2 lines respectively.

For both aluminum oxide and magnesium oxide, any applied stress or residual stress will change the atomic environment of the Cr^{3+} ions. Specifically, the bond lengths between the chromium and oxygen atoms may increase or decrease and this can cause the energy levels involved in the radiative transitions to change. The effect is analogous to that seen for the conduction and valence band edges in a semiconductor subjected to stress. Thus, for a uniform hydrostatic stress, the R_1 and R_2 emission lines of the Cr^{3+} will undergo a uniform downward shift in frequency or energy. On the other hand, for a uniaxial stress, the shift in the R_1 and R_2 lines are different. For compressive uniaxial stress, the two lines actually will approach one another.

For most materials, it is adequate to express the mathematical relationship between the frequency shift, $\Delta \upsilon$ of the luminescence centre emission line and the stress tensor components as [2]:

$$\Delta \upsilon = \pi_{11}\sigma_{11} + \pi_{22}\sigma_{22} + \pi_{33}\sigma_{33} + \pi_{12}\sigma_{12} + \pi_{13}\sigma_{13} + \pi_{23}\sigma_{23} \qquad\qquad (6.6)$$

Here, $\Delta \upsilon = \upsilon - \upsilon_0$ where υ is the frequency of the emission line of the stressed material and υ_0 is the frequency of the unstressed material. σ_{11}, σ_{22}, σ_{33} are the normal stress components of the stress tensor and σ_{12}, σ_{13}, σ_{23} are the shear components of the stress tensor σ_{ij}. The coefficients π_{11}, π_{22}, π_{33}, π_{12}, π_{13}, π_{23} are the corresponding piezospectroscopic coefficients. Thus, the piezospectroscopic coefficient is also a tensor quantity. The relationship between $\Delta \upsilon$ and the stress tensor is linear because the displacement of the atomic positions under stress with respect to the equilibrium position is small compared with the inter-atomic distance. In practice, equation (6.5) can be greatly simplified if one takes advantage of the symmetry of the crystal lattice [2]. When the coordinate axes of the stress tensor are chosen to coincide with the axes of high symmetry of the crystal, some piezospectroscopic components will vanish. In the case of a cubic crystal structure such as MgO, only π_{11}, π_{22}, π_{33} are non-zero and $\pi_{11} = \pi_{22} = \pi_{33}$. Thus, there is only one independent coefficient for this crystal structure. For a crystal structure with trigonal symmetry (*e.g.* alumina) which is less symmetrical than the cubic structure, there are two independent non-zero piezospectrscopic coefficients: $\pi_{11} = \pi_{22}$ and π_{33} with all other coefficients being zero. Munro *et al.* [11] had measured the value of $2\pi_{11} + \pi_{33}$ for the R_1 and R_2 lines of Cr^{3+} in alumina as 7.59cm^{-1} GPa^{-1} and 7.615cm^{-1} GPa^{-1}.

Equation (6.5) above is written in terms of the normal and shear components of the stress tensor σ_{ij}. There is an alternative way of expressing the piezospectroscopic effect and that is to use the hydrostatic and deviatoric components of stress as [2]:

$$\Delta \upsilon = \left(2\pi_{11} + \pi_{33}\right)P + \left(\pi_{33} - \pi_{11}\right)S \tag{6.7}$$

The hydrostatic stress $P = \left(\sigma_{11} + \sigma_{22} + \sigma_{33}\right)/3$ and the deviatoric stress $S = \left(2\sigma_{33} - \sigma_{11} + \sigma_{22}\right)/3$.

Both equations (6.5 and 6.6) show that if the frequency shift of the ruby lines are determined by fitting of the observed spectra, the stress can be deduced using the piezospectroscopic coefficients. If the Young's modulus and Poisson's ratio of the material is known, then this will allow the strain to be determined.

In the above discussion, we have for the sake of clarity limited the explanation of the cause of the piezospectroscopic effect to impurity ions in a material. However, as we will show in section 5, this is by no means the only cause of the piezospectrosopic effect. Recent research has shown that the piezospectroscopic effect is in fact rather general and stress can affect luminescence due to many other physical mechanisms.

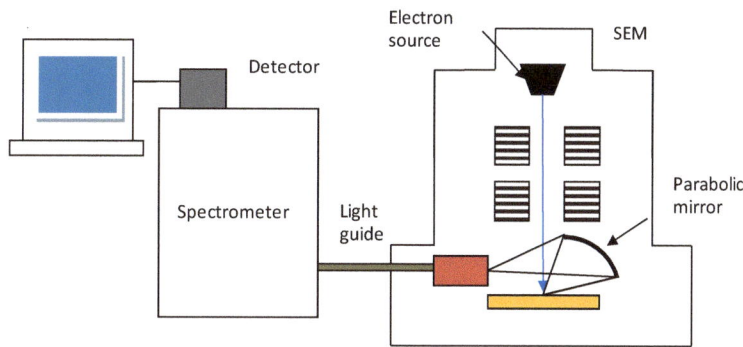

Figure 1: Schematic diagram of a CL measurement system used for stress characterization.

4. EXPERIMENTAL METHOD

The CL method for strain characterization requires an analytical SEM with the necessary attachment to capture the CL excited from the sample. A schematic diagram of a recently typical experimental set-up is shown in Fig. **1**. Before a sample is examined, some sample preparation may be required. For poorly conducting oxides, it is necessary to first deposit a thin layer of carbon to prevent beam charging. Thus, this method is not strictly non-invasive. The electron beam of the SEM excites CL from the sample. In order to minimize the CL generation volume, a lower beam voltage such as 10kV needs to be used. The emitted photons are collected by a paraboloidal mirror and a light guide to a monochromator. Reflective optics are used for light collection to avoid the chromatic aberration due to lenses. Inside the monochromator, a system of mirrors and lenses direct the light to a diffraction grating which disperses the light into its constituent wavelengths. In order to achieve higher throughput, the resulting spectrum is typically recorded by a charge coupled device (CCD) array. However, other detectors such as an optical multichannel analyzer (OMA) can also be used. The OMA consists of a microchannel plate intensifier to amplify the signal and a multi-element photodiode array. By scanning the electron beam under control, a CL spectrum can be obtained for each pixel within the SEM image and one can in this way build up a map of CL spectra with high spatial resolution. In addition to CL spectra, a CL image can also be collected by using a photomultiplier tube (PMT) to collect the CL. Both the broadband CL signal or spectrally resolved CL signal can be used to form a CL image. The CL spectra and/or the CL image can be acquired simultaneously with the secondary electron image or backscattered electrons of the same area of the sample if separate ports are used for these detectors. Such an arrangement can increase the amount of analytical information gathered.

5. APPLICATION EXAMPLES

5.1. Epitaxial III-Nitride Semiconductors

Although the CL technique had been known since the invention of the SEM in the 1940s, it was only relatively recently that this technique was applied to characterize stress in semiconductors. The first detailed study of this kind was performed by Rudloff and co-workers at the Otto-von-Guericke University in Germany in 2003 [7]. This group of investigators used CL to study microcracks in 560 nm thick epitaxial $Al_{0.17}Ga_{0.83}N$ layers grown by metalorganic vapor phase epitaxy (MOVPE) on a (0001) sapphire substrate. Before the $Al_{0.17}Ga_{0.83}N$ growth, a 1.9 μm thick GaN buffer layer was first grown on the sapphire. Due to lattice mismatch between the thick epitaxial layer and the buffer layer, cracks developed on the surface of the $Al_{0.17}Ga_{0.83}N$ to relieve stress. The cracks formed a trigonal network and one had the shape of a parallelogram. This defect was studied in detail by both low temperature CL and room temperature micro-Raman spectroscopy.

The low temperature CL images were collected at different electron beam voltages in order to probe the sample at different depths from the surface (left panels of Fig. 2). The emission wavelength corresponding to the maximum CL intensity is plotted as a function of position to form a CL wavelength image (CLWI) (right panels of Fig. 2). Thus, contrast in the CLWI image indicates a shift in the emission peak of the CL from one pixel to another. At low beam voltages (3kV), there is a noticeable shift to higher energy for the CL peak as the beam is scanned from the middle of the crack free parallelogram towards the cracks on either side. The separation between these two cracks is 13 μm. This blue shift in the CL spectrum is reduced as the beam voltage is increased to 12 kV.

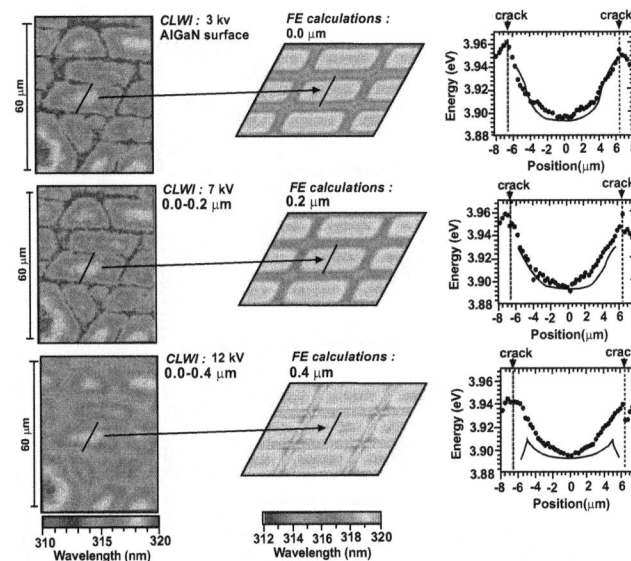

Figure 2: CLWI of $Al_{0.17}Ga_{0.83}N$ film obtained at different electron beam voltage. The stress distribution calculated by finite element simulations is shown in the middle. The CLWI results across a line cross section on the left are compared with results calculated from finite element simulations on the right [7]. Reprinted with permission from D. Rudloff, T. Riemann, J. Christen, Q.K.K. Liu, A. Kaschner, A. Hofmann, Ch. Thomsen, K. Vogeler, M. Diesselberg, S. Einfeldt and D. Hommel, "Stress analysis of $Al_xGa_{1-x}N$ films with microcracks", *Appl. Phys. Lett.*, vol. 82, pp. 367-369, Copyright 2003, American Institute of Physics.

Since the piezospectroscopic coefficients were not known for $Al_{0.17}Ga_{0.83}N$, it was not possible to convert these shifts in the wavelength of the CL emission peak directly into stress. As a consequence, Rubloff *et al.* had to use a simulation approach to investigate the stress pattern in this sample. A finite element simulation based on elasticity theory was applied to an idealized model of the epilayer sample (centre panels of Fig. **2**). This showed that the observed wavelength shift is indeed strongly correlated with the stress across the crack pattern. The inability to deduce stress from the measured CLWI data illustrates the limitation of the piezospectroscopic method, namely, the need to have prior knowledge of the piezospectroscopic coefficients.

5.2. Metal Oxides

As stated in the introduction, Ostertag *et al.* were the first researchers to demonstrate the use of CL to determine residual stresses in strained alumina single crystals. They used a sapphire substrate with an off a-axis (1120) orientation determined by X-ray diffraction. A Vickers hardness indenter made of diamond was used to indent this substrate with a load of 2 N. The indent is pyramidal and the quadrants were oriented approximately parallel to certain symmetry directions of the crystal [2].

Low resolution CL spectra were first measured from the centre of the indent, at the edge of the indent (near a crack) and far from the indent (Fig. **3**). All spectra were collected from an area of 5 μm^2. Two prominent bands were observed. A broad CL band centered at 3.8 eV was found for all three locations and this was attributed to negatively charged oxygen vacancies (F^+ centers). A second broad band at 1.6 eV and the ruby (R) lines were observed only at the edge of the indent and far from the indent. This band is thought to arise from Cr^{3+} ions at low crystal field sites. The quenching of this band at the center of the indent suggested that these ions are absent at this location.

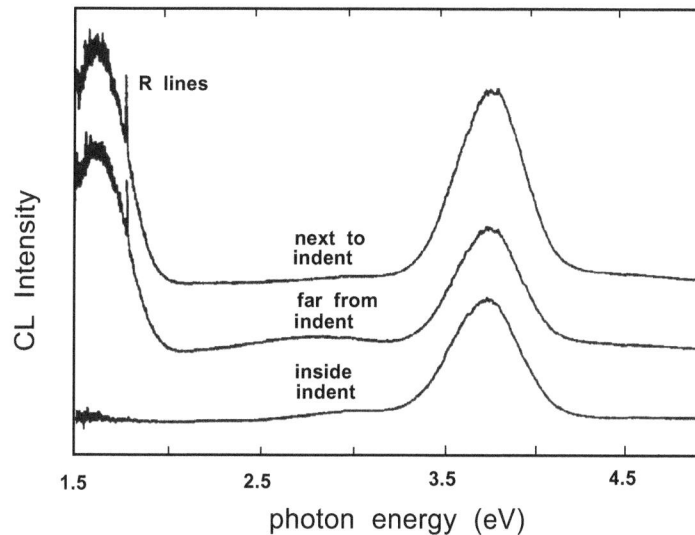

Figure 3: Low resolution CL spectra measured from three regions of sapphire disk [2]. Reprinted from *J. Eur. Ceram. Soc.,* vol. 7, C.P. Ostertag, L.H. Robins and L.P. Cook, "Cathodoluminescence measurement of strained alumina single crystals", pp. 109-116, Copyright 1991, with permission from Elsevier.

Residual stress measurements in the sapphire was carried out using the R_1 and R_2 lines as discussed in section 3 of this chapter. These two lines were resolved in high resolution CL spectra (Fig. **4**). Spectra were collected at the center of the indent, at the edge of the indent and far from the indent. Temperature corrections for the spectra were not applied and the shifts were directly attributed to residual stress. The peak positions of the CL spectra far from the indent were used as the reference because in this location the stress is zero. Only the stress at the edge of the indent could be determined because at the center of the indent, the CL spectrum is completely quenched (Fig. **4**). For the edge of the indent, there is a shift to lower wavenumber (frequency). From the wavelength shifts of the R_1 and R_2 lines, the hydrostatic and nonhydrostatic stress were determined by applying equation (6.6) and the piezocoefficients of Munro [2, 11] and Schawlow and Kaplianskii [2]. Thus,

$$\Delta v_1 = 7.59P - 1.5S \quad \text{(ruby line R}_1\text{)} \tag{6.8}$$

$$\Delta v_2 = 7.615P - 0.6S \quad \text{(ruby line R}_2\text{)} \tag{6.9}$$

Note that since the coefficients of these two simultaneous equations are different, the shifts of the ruby lines are different. By this method, the hydrostatic and deviatoric stresses were found as a function of distance from the center of the indent in three crystallographic directions. The hydrostatic stress is compressive near the centre of the indent and decreased towards zero with distance. The deviatoric stress component is nearly zero in two of the directions. For the third direction, there is a positive deviatoric stress near the center of the indent. The stresses at the center of the indent could not be found.

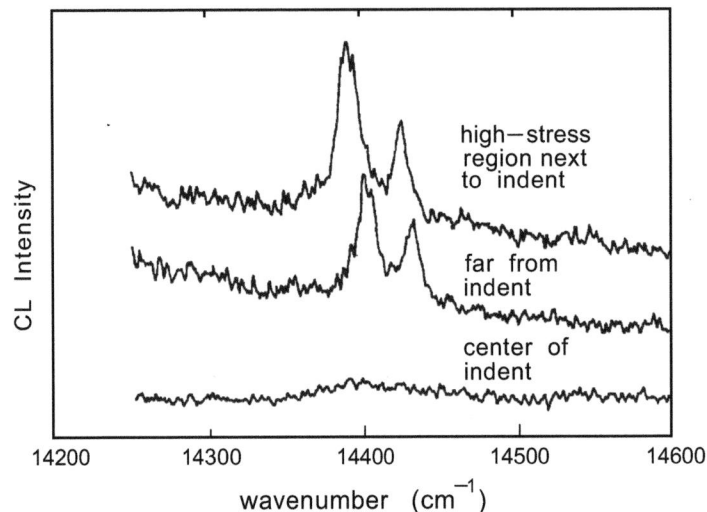

Figure 4: High wavenumber resolution CL spectra of ruby lines taken at three regions of the sapphire disk [2]. Reprinted from *J. Eur. Ceram. Soc.*, vol. 7, C.P. Ostertag, L.H. Robins and L.P. Cook, "Cathodoluminescence measurement of strained alumina single crystals", pp. 109-116, Copyright 1991, with permission from Elsevier.

5.3. Optical Glasses

Residual stresses are important for the fabrication of optical fibers because they can have an adverse impact on the confinement and propagation of light within the fiber core. These stresses are the result of the thermal expansion coefficient mismatch between the fiber core and the cladding. The different viscosities between the two regions can also be a contributory factor. The stress in the fiber core can modify the refractive index through the elasto-optic effect discussed in chapter 3 and increase the critical angle for total internal reflection. Stress can also cause increased Rayleigh scattering which will lead to increased attenuation. Since this is undesirable for optical communications, it is useful to measure the residual stress fields in optical fibers by the cathodoluminescence technique.

Pezzotti and co-workers at the Kyoto Institute of Technology, Japan carried out the first such study using an industrial optical fiber [8]. This group also carried out a study on the stress dependence of the CL spectrum of GaN by a spatially resolved indentation method [10]. The core of this optical fiber has a diameter of 4.8 µm and contained Er_2O_3, Al_2O_3 and GeO_2. As discussed below, these co-dopants enable the piezospectroscopic effect to be applied to yield a stress map of the fiber core region. The cladding of the fiber is a high silica glass with unknown impurities and has a diameter of 120 µm. A short section of this fiber was cut perpendicular to the fiber axis and is embedded within an epoxy resin. The epoxy was polished by fine diamond paste and further processed by low angle ion beam milling to reduce the residual stress due to the machining.

The CL measurements made use of a SEM with a thermal field emission gun. The sample chamber was fitted with a micro-jig and a load cell. These two components together allow the piezospectroscopic coefficients to be determined from the CL spectra. Although the beam size of was 1.5nm, the actual spatial resolution was only about 3nm due to beam spreading. This was achieved only for low beam voltage (1 kV) and high beam current (80 μA).

CL spectra were collected from both the fiber core and the cladding (Fig. **5**). The spectra from the fiber core consist of two prominent peaks. However, it was found that for reproducible fitting, it is necessary to use four hybrid Gaussian/Lorentzian functions as described in section 3. These functions have peaks at 410 nm, 460 nm, 630 nm and 650 nm. The 410 nm peak is due to di-coordinated Ge within the glass [8]. The 460 nm peak is caused by oxygen deficiency in the silica structure while the 630 nm peak is due to excess of oxygen. Both these peaks are extremely useful because they have piezospectroscopic properties in the visible region of the spectrum like the Cr^{3+} ions. In other words for glasses, it is not even necessary to have impurities in the glass. This property is useful for interconnect stress characterization in the next section.

The CL spectrum from the cladding of the fiber consists of only the 460nm and 630nm peaks. By measuring the CL spectra from the fiber at different applied loads, the piezospectrosopic coefficient for each peak can be deduced. This calibration involves essentially two steps. First, beam theory is applied to calculate the stress at each applied force from the load cell. Secondly, the peak positions of the CL spectra are determined by mathematical fitting and then the shift is calculated by reference to the peak position of the unstressed material. In order to ensure zero stress, the material cut from the fiber is ground to a fine powder and no external stress is applied during the CL measurement of the powder. The piezospectrosopic coefficients, Π_u for the 410 nm, 460 nm and 630 nm peaks are determined thus to be: 6.744 nm/GPa, -6.529 nm/GPa and 7.137 nm/GPa respectively. These coefficients are for uniaxial stressing and it is assumed that for an amorphous solid, $\pi_{11} = \pi_{22} = \pi_{33}$.

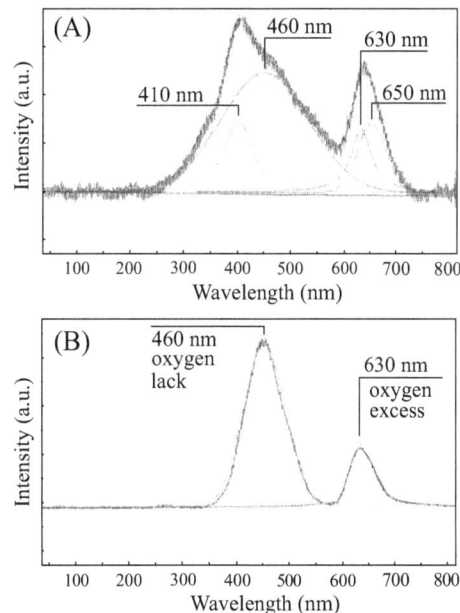

Figure 5: CL spectra taken from the core region (A) and the cladding region (B) of an erbium doped optical fiber. Reprinted with permission 'Piezo-spectroscopic assessment of nanoscopic residual stresses in Er^{3+}-doped optical fibres', Giuseppe Pezzotti, Andrea Leto, Katsuhisa Tanaka and Orfeo Sbaizero, J. Phys.: Condens. Matter 15, 7687-7695, November 2003, IOP Publishing.

SEM and CL images were simultaneously taken at a location near the core cladding interface. For this location, the intensity of the 410nm band was used to generate the CL image. Although the contrast between the core and cladding was very weak in the secondary electron image (SEI), the CL image showed a clear interface demarcating the core and cladding. This is because Ge is only present in the core region and only a small amount of Ge diffused into the cladding near the interface during the fabrication of the fiber. Since the piezospectroscopic coefficient for the 410nm band is known, a spatially resolved hydrostatic stress map can be extracted from the array of CL spectra collected. The stress in the core is tensile and is highly inhomogeneous micelle-like. On the other hand, the stress in the cladding is compressive. Pezotti *et al.* argued that this could be due to either nanocracks or the co-existence of a strong and weak glass network.

A similar set of SEI, CL and stress images were collected at the cladding. For this region, the 460nm band had to be used. Cracks could be seen in the SEI and these correspond to regions of tensile stress in the stress image. The ability to detect these small cracks is very useful to industry because these cracks are a potential source reliability issues.

5.4. Interlayer Dielectrics

Residual stresses are also of significance to the dielectric layers in the interconnect stacks of integrated circuits. This is especially the case for the low-*k* and porous low-*k* dielectrics that are integrated with copper metallization nowadays. Excessive stresses can lead to delamination at interfaces and time dependent dielectric breakdown reliability problems. These stresses can be due to mismatch in the thermal expansion coefficients of the different interconnect materials or intrinsic stresses. In 2005, Kodera *et al.* of Toshiba corporation applied the CL method to detect residual stress in patterned copper SiO_2 interlayer dielectrics (ILD) [9]. They applied the finding of Pezzotti about the defect related luminescence in SiO_2 in [8] and were able to observe CL from the ILD despite the fact that the SiO_2 does not contain metallic luminescence centers. This is because metallic impurities can diffuse under electric field and can cause serious reliability problems in integrated circuits. The trace of the piezospectroscopic coefficients was calibrated by using an indenter to make a crack in the SiO_2 and a CL spectrum was taken next to the crack.

Both stress as a function of linear position and stress maps in tetraethyl orthosilicate (TEOS) low-*k* films were reported. For these measurements, damascene Cu/TEOS interconnect samples were first fabricated. Then by applying the measured pizospectrosocpic coefficients, the stress as a function of distance from the Cu conductor was measured. A tensile stress was found in this way in the vicinity of the Cu conductor. This stress could be determined even in the case of narrowly spaced Cu conductors because of the high spatial resolution of the CL technique.

5.5. Piezospectroscopic Coefficients of GaN

The calibration technique mentioned in the previous section was applied to bulk (0001) Si: GaN crystals [10]. It is important to find out the piezospectroscopic coefficients for GaN because this wide band gap semiconductor has great technological significance for temperature tolerant field effect transistors and white light emitting diodes for solid state lighting. One of the main challenges is the growth of high quality epitaxial layers on non-sapphire substrates and a detailed knowledge of the stress fields around defects is thus critical to solving this challenging crystal growth problem [10]. If the piezospectrosocpic coefficients are unknown, then as in section 1, it would be impossible to extract stress field information from the measured CL spectra at different spatial locations. By using a diamond tip indenter, indentations were made on the surface of a GaN crystal and the residual stress at the crack tip was probed by CL in a FE-SEM. For GaN, the luminescence originated from an excitonic band. From the shift of the luminescence peak, the magnitude of the piezospectroscopic coefficient was found to be 1.35 nm/GPa. In obtaining this value, it was necessary to use a pre-determined probe response function.

REFERENCES

[1]   S.E. Molis and D.R. Clarke, "Measurement of stresses using fluorescence in an optical microprobe: stresses around indentations in a chromium-doped sapphire," *J. Am. Ceram. Soc.*, vol., 73, pp. 3189-3194, Nov. 1990.

[2]   C.P. Ostertag, L.H. Robins and L.P. Cook, "Cathodoluminescence measurement of strained alumina single crystals," *J. Eur. Ceram. Soc.*, vol. 7, pp. 109-116, Mar. 1991.

[3]   A. Kitai, *Luminescent materials and applications.* John Wiley & Sons, Ltd.: Chichester, 2008.

[4] R.I. Todd, D. Stowe, S. Galloway, D. Barnes and P.R. Wilshaw, "Piezospectroscopic measurement of the stress field around an indentation crack tip in ruby using SEM cathodoluminescence," *J. Eur. Cer. Soc.,* vol. 28, pp. 2049-2055, Jul. 2008.

[5] B.G. Yaccobi and D.B. Holt, "Cathodoluminescence scanning electron microscopy of semiconductors," *J. Appl. Phys.,* vol. 58, pp. R1-R24, Feb. 1986.

[6] L. Grabner, "Spectroscopic technique for the measurement of residual stress in sintered Al_2O_3," *J. Appl. Phys.,* vol. 49, pp. 580-583, Feb. 1978.

[7] D. Rudloff, T. Riemann, J. Christen, Q.K.K. Liu, A. Kaschner, A. Hofmann, Ch. Thomsen, K. Vogeler, M. Diesselberg, S. Einfeldt and D. Hommel, "Stress analysis of $Al_xGa_{1-x}N$ films with microcracks," *Appl. Phys. Lett.,* vol. 82, pp. 367-369, Jan. 2003.

[8] G. Pezzotti, A. Leto, K. Tanaka and O. Sbaizero, "Piezo-spectroscopic assessment of nanoscopic residual stresses in Er^{3+}-doped optical fibres," *J. Phys. Cond. Mat.,* vol. 15, pp. 7687-7695, Nov. 2003.

[9] M. Kodera, S. Uekusa, S. Kakinuma, Y. Saijo, A. Fukunaga, M. Tsujimura and G. Pezzotti, "Nanometre-scale stress detection of patterned ILD using cathodoluminescence piezo-spectroscopic assessments in a nano-stress microscope," In: *Advanced metallization conference Asian session,* 2005, pp. 58-59.

[10] A.P. Porporati, Y. Tanaka, A. Matsutani, W. Zhu and G. Pezzotti, "Stress dependence of the near-band-gap cathodoluminescence spectrum of GaN determined by spatially resolved indentation method," *J. Appl. Phys. vol.,* 100, pp. 083515-1-7, Oct. 2006.

[11] R.G. Munro, G.J. Piermarini, S. Block and W. B. Holzapfel, "Model line-shape analysis for the ruby R lines used for pressure measurement," *J. Appl. Phys.* vol. 57, pp. 165169 Jan. 1985.

CHAPTER 7

Nano-Beam Diffraction and Convergent Beam Electron Diffraction

Abstract Two electron crystallographic techniques capable of measuring strain in nanoscale regions of a device sample are discussed. Both require the use of an analytical TEM. In nano-beam diffraction, a small collimated beam is diffracted by a crystalline sample and by measuring the distances between diffraction spots, the lattice spacing in the strained region can be found. This together with a separate measurement at the unstrained region will allow the strain to be characterized. The second method uses a convergent electron beam and involves measuring the shifts in the higher order Laue zone lines which are more sensitive to strain. In order to access these lines in the analytical TEM, sample tilting is required. This however limits the spatial resolution of the convergent beam electron diffraction technique to about 50nm.

Keywords: Electron diffraction, Nanobeam, Convergent beam, Bragg law, Reciprocal lattice.

1. INTRODUCTION

In this chapter, we will discuss two strain characterization techniques that are applied to individual semiconductor devices because of their nanoscale spatial resolution. Both these techniques make use of the wave nature of the electron and the fact that its de-Broglie wavelength can be comparable to the inter-atomic spacing in a semiconductor. When an electron beam with high energy is transmitted through a thinned crystalline specimen, the electron wave will diffract and superpose to form a diffraction pattern characteristic of the sample. Since this diffraction pattern is dependent on the lattice structure, it can be used to measure the semiconductor lattice spacing [1]. Two electron diffraction techniques have been used in recent years: nano-beam electron diffraction (NBD) and convergent beam electron diffraction technique (CBED). Unlike the methods of earlier chapters, the strain can be calculated or simulated from first principles using geometric measurements (lengths and angles). Knowledge of material properties is not required and as such can be applied to any crystalline sample.

2. NANO-BEAM ELECTRON DIFFRACTION

The NBD technique is the more recent electron diffraction technique. It was developed by K. Usuda and co-workers at the National Institute of Advanced Industrial Science and Technology (AIST) in Japan in 2004 specifically for studying the strain distribution within strained silicon devices [2].

2.1. Principle of Electron Diffraction

Electron diffraction is one of the key experiments that demonstrated the wave-particle duality of matter [3]. Electrons can diffract from a crystalline solid because for atomic dimensions, the motion of an electron cannot be precisely predicted. Instead, one can only know precisely the probability density of the electron. This probability density gives the chance of finding an electron at a particular region of space and is obtained from the wavefunction of the electron. Inside an electron microscope, the wavelength λ of this electron wavefunction is given by [4]:

$$\lambda = \frac{12.2643}{\sqrt{\left(E_0 + 0.97845 \times 10^{-6} E_0^2\right)}} \tag{7.1}$$

where E_0 is the electron accelerating voltage in volts and the wavelength is in the unit of Angstrom. When the electrons in an electron microscope enter a crystalline sample, they will interact with the valence electrons of the atoms in the crystal. Since both are of the same charge, the incident electrons will be scattered (deflected from the original path). For elastic scattering, there is no change in the wavelength of the electron and only the direction of propagation is changed. In many directions, the wavefunction of the electrons scattered by atoms in different lattice planes of the crystal will be out of phase and they will simply cancel. However, in certain specific directions, the phase of the electron wavefunctions will be offset by integer multiples of 2π radians and thus will be in phase. This is expressed by the Bragg condition [3]:

Terence K.S. Wong

$$n\lambda = 2d_{hkl} \sin\theta_B \qquad\qquad (7.2)$$

Here, θ_B is the Bragg angle and d_{hkl} is the inter-planar spacing of the lattice plane indexed by the Miller index *hkl*. (Fig. **1**) This equation which is also applicable to X-ray diffraction in chapter 11, can be used to find the inter-planar spacing if the electron wavelength and the Bragg angle are known. Note that for small angle diffraction, we can make a small angle approximation and write equation (7.2) as:

$$n\lambda = d_{hkl}\Theta \qquad\qquad (7.3)$$

Here $\Theta = 2\theta_B$, the total scattering angle is the angle between the direction of the incident electron and the diffracted electron.

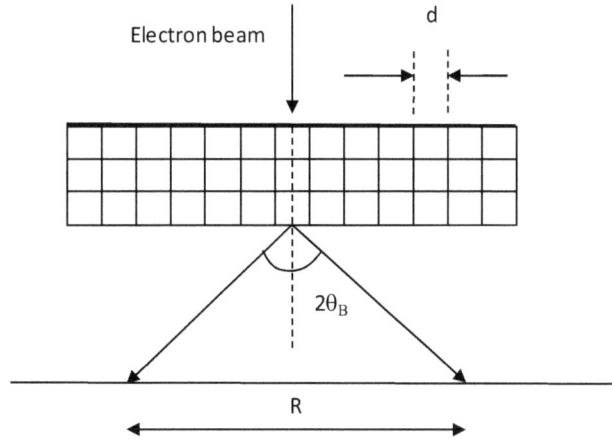

Figure 1: Schematic diagram illustrating the principle of NBD.

2.2. Experimental Technique of NBD

The NBD strain measurement technique is performed within a transmission electron microscope (TEM). An electron beam with a diameter as small as 10nm is incident on the sample and is transmitted through a thinned specimen. Since the technique is especially useful for device structures, the thinning technique employed is focused ion beam (FIB) milling. After locating the feature of interest, an ion beam is used to physically remove atoms from the area to be observed by the electron beam. During milling, care must be exercised to ensure that the sample is not so thin that strain relaxation occurs.

After sample preparation, the specimen is analyzed in a TEM by recording a diffraction pattern of the incident electron beam. If the incident beam is of small diameter, it is possible to obtain diffraction patterns from multiple points within the same area of the sample. An important aspect of the NBD technique is that it only detects relative changes in the lattice spacing of a crystalline sample. Thus, if strain is the quantity to be measured, then one must also measure the NBD pattern of a unstressed semiconductor sample which is the reference sample. Furthermore, since the beam is incident on a thin two dimensional specimen, the NBD technique allows one to determine the change in lattice spacing in two spatial directions. For a CMOS device with the classical structure, this means the lattice deformation in the direction of current flow and the deformation perpendicular to current flow can be deduced. Once the diffraction patterns are recorded, a numerical fitting is performed to find the diffraction peak positions.

2.3. NBD Characterization of Strained Silicon Substrates

The NBD technique was first applied to strain characterization in strained silicon substrates by Usuda and co-workers in 2003 [5]. Two types of strained silicon substrates were studied. In the first, the strained silicon layer was grown on a relaxed SiGe on insulator substrate. This type of substrate is also known as a

SGOI substrate. SGOI combines the advantages of higher carrier mobility in the strained silicon and the lower parasitic capacitance of the buried oxide substrate.

A schematic diagram of this substrate is shown in Fig. **2a**. First, SiGe is grown epitaxially by ultra high vacuum chemical vapor deposition (UHV-CVD) on a silicon substrate by using a mixture of disilane (Si_2H_6) and germane (GeH_4) gases. After growth, a buried silicon oxide layer was formed by a high dose high energy oxygen implant into the silicon substrate. In order to form the SiO_2, a high temperature post implant anneal was performed. During this anneal, some oxidation within the SiGe also took place. As a result of this so called internal thermal oxidation (ITOX) [5], Ge condensation takes place and a SiGe segregates from the SiO_2 to form a trilayer structure of SiO_2/SiGe/buried oxide. After annealing, a chemical dry etch (CDE) was performed after lithography to create islands of relaxed SiGe on the buried oxide. Silicon is then deposited by CVD onto the patterned substrate. The thin silicon layer deposited onto the relaxed SiGe becomes strained because of the larger lattice parameter of the relaxed SiGe.

After cross sectioning, NBD patterns were obtained in the centre part of the strained silicon layer, the SiGe relaxed layer and the silicon substrate. The beam size is about 10 nm [5]. These locations are indicated by a cross in Fig. **2a**. The latter is necessary because only relative lattice parameter changes can be deduced from NBD. By assuming that the lattice parameter of the silicon substrate is 0.543 nm, the lattice parameter of the SiGe layer was found to range from 0.546 nm – 0.552 nm. This is in agreement with the lattice parameter calculated from Vegard's law using the Ge concentration determined by SIMS. For the strained silicon layer, the electron diffraction pattern showed that the lattice has undergone a tetragonal deformation and the layer is clearly under tensile strain. This demonstrates the ability of NBD to measure changes in lattice parameters in two orthogonal directions. During CVD, strained silicon is also formed on the edge of the SiGe mesa. When the strained silicon near the edge of the SiGe mesa structure was measured by NBD, some strain could be detected from the diffraction image. However, the strained silicon is more relaxed than that near the centre part of the mesa.

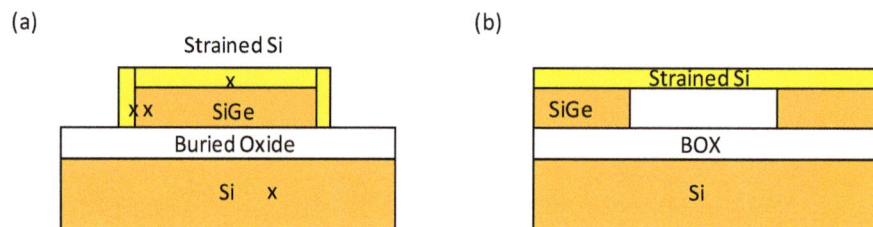

Figure 2: Schematic diagram of (a) the SGOI substrate and (b) SSON substrate.

In actual strained silicon devices on SGOI substrates, isolation structures must be built around each device. Due to ULSI scaling, the separation distance between the device and the isolation structure is expected to be small. Since it is known that the stress field from the isolation structure can affect the characteristic of silicon devices, a further study was carried out to investigate the effect of the isolation oxide [2]. The strained silicon on SGOI was prepared by the Ge condensation process as described above. The thickness of the strained silicon, SiGe and buried oxide after the ITOX were 14nm, 40nm and 100nm respectively. After the strained silicon on SGOI layers were etched to 5μm mesas by CDE, a layer of CVD SiO_2 was deposited to simulate the effect of the isolation oxide. These isolated strained silicon mesas were then annealed by rapid thermal annealing in a nitrogen ambient at four maximum temperatures: 700°C, 800°C, 900°C and 1000° C for 30s. The ramp rates during heating and cooling were 450° C/min and 150° C/min respectively. For comparison, samples which were not subjected to mesa etching and isolation oxide were similarly annealed. When the Raman spectra of both sets of samples were compared, it was found that a small amount of relaxation (~3%) of the strain in the strained silicon had occurred for the mesa samples. There could be two reasons for this. First, there might have been some Ge diffusion into the strained silicon or the relaxation could be due to the isolation oxide. The first possible cause was however, not consistent with SIMS analysis and the Raman spectra. As a confirmation, more precise measurements by NBD was performed for each layer of the mesa samples (strained silicon, SiGe and silicon substrate) after the rapid thermal annealing step (Fig. **3**). From the electron diffraction patterns, the lattice parameters in the in plane and out of plane directions were deduced. The strained silicon layer is still tensile strained in the in plane direction. The lattice parameters for the SiGe layer are consistent with a relaxed layer. Thus, the conclusion is that the presence of the isolation oxide has a minimal effect on the strain in the strained silicon on SGOI substrates.

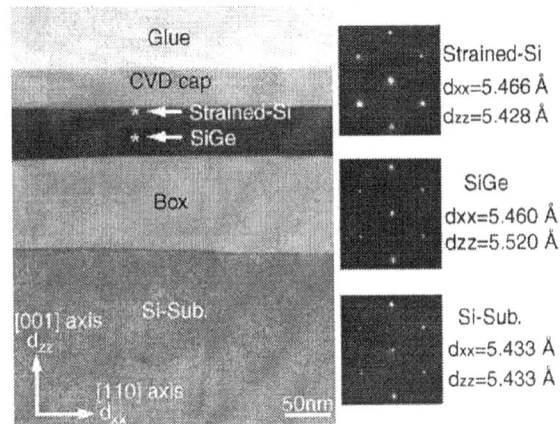

Figure 3: NBD pattern measured at the strained silicon layer, SiGe layer and the silicon substrate [2]. Reprinted from *Appl. Surf. Sci.,* vol. 224, K. Usuda, T. Mizuno, T. Tezuka, N. Sugiyama, Y. Moriyama, S. Nakahari and S. Takagi, "Strain relaxation of strained-Si layers on SiGe-on-insulator (SGOI) structures after mesa isolation", pp. 113-116, Copyright 2004, with permission from Elsevier.

2.4. NBD Characterization of Strained Silicon On Nothing

The second type of strained silicon substrate investigated was strained silicon on nothing (SSON) (Fig. **2b**). This provides the lowest possible parasitic capacitance and is similar in principle to the air gap structures found in advanced copper interconnects. The process steps are the same as the SGOI substrate up to the post oxygen implant anneal step. For SSON, there is no mesa etching. Instead, the strained silicon is deposited onto an un-patterned SiGe surface. This is followed by a selective etch of the SiGe beneath the strained silicon to form a freestanding thin strained silicon layer on air (hence the term nothing) [5]. From cross sectional TEM, it was seen that the strained silicon layer was indeed separated from the substrate and was supported only on two sides by the SiGe piers. NBD was used to characterize the strain in the middle of the free standing strained silicon. The lattice parameter in the direction parallel to the interface was found to be 0.545nm which is very slightly larger than the lattice parameter of the silicon substrate. The strain in the strained silicon near the SiGe piers was also measured by NBD. These were found to be greater and thus, there is some strain relaxation in the strained silicon at the centre.

2.5. NBD Characterization of SGOI MOSFET Channels

In a subsequent study by the same research group [6], the high spatial resolution advantage of NBD was used to study the strain distribution within strained silicon MOSFET channels on SGOI substrates. The SGOI substrates were fabricated by the Ge condensation technique [6] applied to conventional SOI substrates. SiGe was grown by UHV-CVD. The advantage of the Ge condensation method is that a high Ge concentration can be realized in the SiGe layer without the need to grow a thick layer as in the conventional graded layer approach. In this case, the SiGe layer was 84nm and the buried oxide layer was 100nm thick. A final layer of strained silicon 24nm was grown on the SiGe.

Strained silicon MOSFETs were fabricated on the SGOI substrates and NBD was used to measure the strain at discrete locations within the channel and outside the channel. Fig. **4** shows a MOSFET with 35nm gate length. The white dots indicate the positions where NBD measurements were made. The strain in the x-direction deduced from NBD is tensile and is relatively uniform (0.6%). On the other hand, the strain in the z-direction (perpendicular to channel) is compressive. The middle of the channel is slightly more compressive (-0.25%) than near the source and drain.

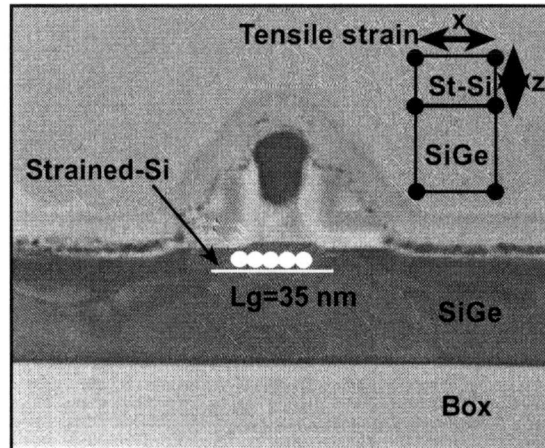

Figure 4: Cross sectional TEM micrograph of a 35nm gate length strained silicon MOSFET. The white dots indicate the locations where NBD strain measurements were made [6]. Reprinted from *Mater. Sci. Eng. B*, vol. 124-125, K. Usuda, T. Numata, T. Irisawa, N. Hirashita and S. Takagi, "Strain characterization in SOI and strained-Si on SGOI MOSFET channel using nano-beam electron diffraction (NBD)", pp. 143-147, Copyright 2005 with permission from Elsevier.

3. CONVERGENT BEAM ELECTRON DIFFRACTION

Convergent beam electron diffraction (CBED) is a related electron microdiffraction technique that had been applied for characterizing strain in semiconductor samples. Like NBD, it can yield strain information from a small area of the sample. The area is limited by the probe size as well as the sample orientation. CBED is usually carried out in an analytical TEM because it uses a focused electron probe to study the sample. As the name implies, a convergent electron beam from the electron optical column of the TEM is incident on a crystalline sample. This is the main difference from NBD where a collimated or parallel beam is used. Due to the periodicity of the sample atoms, electron diffraction occurs within the sample and the diffracted beams have characteristic angles with respect to the incident beam. The use of a convergent beam causes the characteristic diffraction directions to appear as lines (see below) in the CBED image. It is those lines corresponding to the higher order Laue zones (HOLZ) that are used for strain measurement. These HOLZ lines are used because they correspond to higher order reciprocal lattice vectors of the crystal and are more sensitive to small changes in the lattice dimensions.

3.1. Principle of CBED

The principle of CBED is discussed thoroughly in the text by Spence and Zuo [4]. Let us consider the simplest case of a single set of crystal planes in a thin sample oriented perpendicular to an incident electron beam. Suppose initially, the beam is also parallel as in NBD. Then according to the Bragg equation, this configuration will give rise to a set of small diffraction spots. Now suppose the condenser lens system of the TEM is used to form a small probe at the sample surface (Fig. **5**). This change to a convergent beam will cause the diffraction spots to broaden into several larger disks arranged side by side. It is not difficult to see why this is the case. Consider the central disk. This disk corresponds to those electrons which have not been deflected from their incident path. For each point in the central disk, one can extrapolate back to the final condenser lens aperture and find a corresponding point source. The condenser lens aperture can thus be thought of as a collection of incoherent point sources. Each such point source will give rise to a set of diffracted points satisfying the Bragg equation. In other words, the disks referred above represent an ensemble of point diffraction patterns laid out side by side. The distance between adjacent points, X is approximately proportional to the reciprocal lattice vector, g [4]:

$$X = Lg\lambda \tag{7.4}$$

where L is the distance between the detector and the sample.

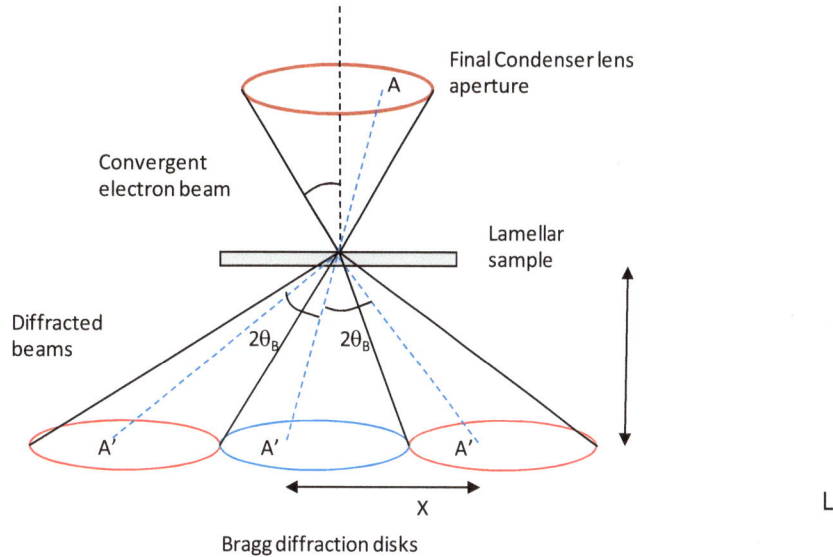

Figure 5: Schematic diagram of the principle of convergent beam electron diffraction.

Before proceeding further, it is necessary to define two terms in reciprocal space. One of the basic concepts in solid state physics is that for every crystal lattice in real space, one can use the lattice vectors to compute a corresponding lattice in reciprocal space called the reciprocal lattice [7]. This is because the unit of magnitude in this lattice is the inverse cm. Like the real space lattice, the reciprocal lattice is made up of lattice points spanned by unit reciprocal lattice vectors. A plane of reciprocal lattice points passing through the origin of the reciprocal lattice is called the zero order Laue zone or ZOLZ [4]. After identifying one such ZOLZ, one can immediately recognize the higher order Laue zones (HOLZ) which are any other plane of lattice points parallel to the ZOLZ that, however, does not contain the origin of the reciprocal lattice. Since the HOLZ can comprise a number of planes, these are also sometimes called the first order Laue zone (FOLZ) ($n = 1$), second order Laue zone (SOLZ) ($n = 2$) and third order Laue zone (TOLZ) ($n = 3$) and so forth.

By using the Ewald sphere method [7], one can work out which reciprocal lattice point will fulfill the Bragg condition for electron diffraction. Suppose one such point in one of the HOLZ planes is found and this has the reciprocal lattice vector g. By considering the Bragg condition, we can see that there should be a cone of possible orientations of the incident wave vector K that satisfy the Bragg equation. Similarly, there is a cone of diffracted wave vectors K' that forms a mirror image of the incident cone. The magnitude of the wave vector K or K' is the inverse of the electron wavelength. For high energy electrons, the electron wavelength is very short and therefore the cone radius in reciprocal space will be very large. The projection of these cones onto the observation plane thus approximates to a straight line and these are called the HOLZ lines (Fig. **6**). One definition of the HOLZ lines is that they are the locus of points satisfying the Bragg equation for a higher order reciprocal lattice vector, *g*. Note that HOLZ lines occur in pairs: one for the incident ray in the central disk and one for the diffracted ray in another disk. The reader should now realize why a convergent beam is used in CBED. The convergent beam is basically a cone of incident electrons which after higher order diffraction from a set of crystallographic planes will give rise to HOLZ lines in the observation plane.

The location of HOLZ lines within say the central disk is governed by Bragg's law and thus depends on the spacing of the lattice planes. These lines are therefore sensitive to the changes in the lattice spacing and can be used to determine strain. For an inter-planar spacing of *d*, the strain is written as $-\Delta d/d$ and this is given by [4]:

$$-\frac{\Delta d}{d} = \frac{\Delta \theta}{\theta} = \frac{\Delta g}{g}$$

(7.5)

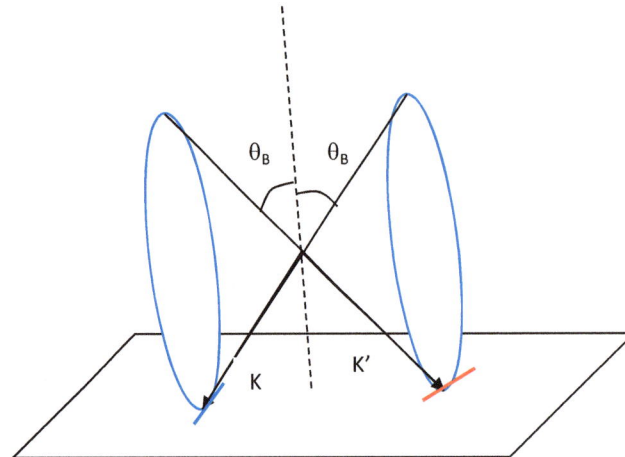

Figure 6: Schematic diagram illustrating the origin of HOLZ lines in CBED measurements.

The sensitivity to strain increases with the diffraction angle and is the reason why HOLZ lines are preferred. In order to determine the change in the reciprocal lattice vector, the intersections of HOLZ lines can be used.

3.2. CBED Experimental

An analytical TEM is needed for CBED characterization of lattice strain. There are two basic steps involved: (i) focusing the beam onto the sample to form a nanoscale probe and (ii) obtaining the CBED pattern in the imaging mode. First, the probe is focused at an unstrained region of the sample and a CBED pattern is recorded. This step is needed to determine the effective beam voltage. One needs to know the lattice spacing of the unstrained region and set the beam voltage as the fitting parameter. Next, the CBED pattern of the strained region is recorded and the effective beam voltage is set as a fixed parameter and the lattice parameters are adjusted by software to find the best fit to the experimental CBED pattern.

The TEM should be equipped for CBED measurements and these include allowing the sample to be tilted with respect to the beam while maintaining a small probe size [4]. The emission intensity of the electron gun should be stable and contamination should not build up during the measurement. It is therefore desirable to have a sample holder that can cool the sample because this makes it easier to view the HOLZ lines. Another important hardware is the energy filter before the detector. This is for avoiding the inelastically scattered electrons and to improve the CBED pattern. In addition to microscope hardware, CBED analysis software available commercially is also essential (www.soft-imaging.net).

3.3 Application Examples

A. Locally Oxidized Silicon Isolation Structures

One of the first applications of CBED is the characterization of strain in locally oxidized silicon isolation structures (LOCOS) in silicon large scale integrated circuits (LSI). This was the most widely used isolation technique before shallow trench isolation (STI) was introduced. Isolation is achieved by the oxidation of a thick field oxide layer while the active regions of the wafer are prevented from oxidation by an oxide/nitride bilayer. Due to the different thermal and mechanical properties of the dielectrics and silicon, stress can build up in the silicon as a result of the LOCOS process and this can lead to unwanted dislocations and defects. As mentioned in chapter 5, these isolation structures were also studied by the micro-Raman technique.

In 1993, Kimoto and co-workers at the Hitachi central research laboratory applied the CBED technique to measure the strain distribution in a LOCOS structure [8]. The 2 mm x 2 mm sample used was cut from a (001) LOCOS wafer with a 700nm oxide film. The wafer was first thinned mechanically to 20 μm and then milled by Ar^+ ions to a final thickness of 400 nm. CBED was performed using a 100 kV field emission TEM with an electron beam spot was 8 nm. The positions of the HOLZ lines were measured from the negative of photographic films using a projector.

The CBED pattern of unstrained silicon (001) was first obtained and this was used to determine the effective acceleration voltage. Using a kinematic approximation, the HOLZ lines can be calculated by using the known lattice parameter of silicon and the acceleration voltage which is treated as a fitting parameter. The acceleration voltage that best reproduce the experimental CBED pattern is the effective acceleration voltage.

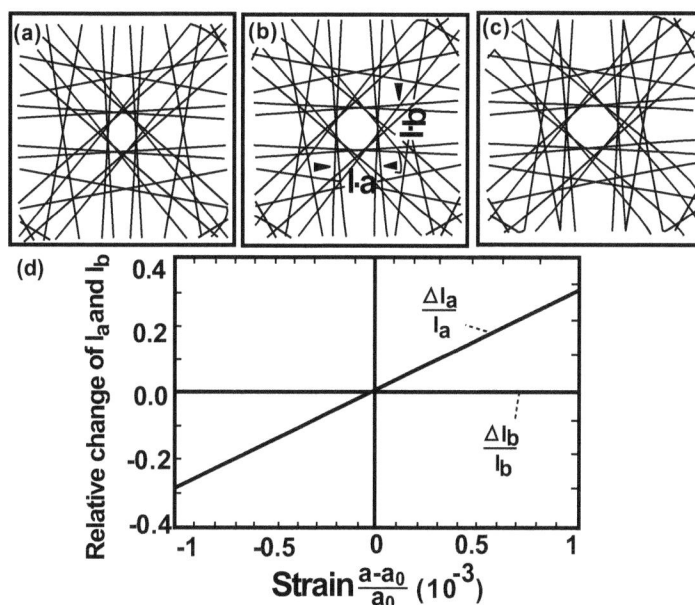

Figure 7: Simulated HOLZ line patterns and the linear relationship between the lattice strain of silicon and relative change in HOLZ line separation [8]. Reprinted from K. Kimoto, K. Usami, H. Sakata and M. Tanaka, "Measurement of strain in locally oxidized silicon using convergent-beam electron diffraction", *Jpn, J. Appl. Phys.,* vol. 32, pp. L211-L213, 1993 with permission from Japanese Society of Applied Physics.

If uniaxial strain along the [100] direction of silicon is considered, the simulated CBED pattern using the effective beam voltage will be slightly different because of changes in the lattice parameters (Fig. **7**). These changes can be related to the shifts in the positions of the HOLZ lines in the CBED pattern. Kimoto *et al.* used the distance between the intersection points of four HOLZ lines designated as (17, 1, 1) and (17, -1, 1) and (-17, 1 ,1), (-17, -1, 1) [8]. They found that the fractional change in this distance l_a is linearly related to the uniaxial strain in the silicon.

$$\frac{a-a_0}{a_0} = 3.5 \times 10^{-3} \frac{\Delta l_a}{l_a} \qquad (7.6)$$

a_0 is the lattice parameter of unstrained silicon and a is the lattice parameter after uniaxial strain. Using this empirical relationship, the CBED pattern of nine points along a line perpendicular to the edge of the LOCOS oxide was measured. For each CBED pattern, the distance l_a was measured and the change Δl_a was determined by using the l_a of the CBED pattern taken at a point 30μm from the oxide edge. When the strain $(a-a_0)/a_0$ was plotted against the distance from the oxide edge, the strain was found to be compressive and had a minimum of 4×10^{-4} at about 1mm from the edge of the silicon oxide. The accuracy of this strain measurement depends on the value of l_a. Several measurements of l_a at different points in the unstrained silicon was made and from the experimental uncertainty the minimum detectable strain was found to be 1×10^{-4}. Thus, the strain measured at different distances from the silicon should be within experimental error.

B. Shallow Trench Isolation Structures

Shallow trench isolation (STI) structures are used for electrically separating transistors and other front end devices on a substrate. For dynamic random access memory (DRAM) manufacturing, STI is usually the first step. The process involves depositing an insulator such as silicon oxide into a shallow trench etched in silicon. Since stress is induced in silicon by the silicon oxide, dislocations can be generated in the surrounding silicon and these provide leakage paths that are detrimental to DRAM devices. Local strain measurement with high spatial resolution is one way of monitoring this reliability problem in state of the art DRAMs.

High resolution strain measurement of STI structures by CBED was demonstrated by Kim and Park of the Samsung Advanced Institute of Technology (Suwon, Korea) in 2004 [9]. The STI test structure studied was filled with a polysilazane based spin on glass (P-SOG). This inorganic spin on dielectric is of interest because it had shown less leakage than oxide deposited by high density plasma deposition (HDP).

The CBED patterns were obtained using a FEI Tecnai UT-30 TEM operating in the STEM mode. The beam voltage for the experiment was 300 kV and the probe diameter was 2 nm. A cross sectioned sample with both P-SOG and silicon and a thickness of 200 nm was used. CBED patterns at four separate locations were recorded by tilting the sample by 11.3° towards the [230] zone axis. One location is in the substrate and the rest are near the dielectric. For each location, a portion of the recorded CBED pattern was used for strain determination. The strain was deduced by matching a simulated HOLZ pattern to the measured HOLZ pattern. An example of this is shown in Fig. **8**. The pattern matching is performed by using the program EXTAL [9] for general electron crystallography. The correlation between the experimental and theoretical patterns is calculated by [9]:

$$XCF = \sum_i I_i^{theo} I_i^{expt} (p_1....p_n) \tag{7.7}$$

In the cross correlation function XCF, I_i^{theo} and I_i^{expt} are respectively the theoretical and experimental intensities over the ith pixel; $p_{1...} p_n$ are the fitting parameters and can be quantities like the sample thickness or lattice parameter. For this particular study, only three parameters were allowed to be adjustable. By performing pattern matching at all four locations, it was found that the P-SOG filling does reduce the strain in the silicon relative to the HDP silicon oxide. The accuracy of the deduced strain was found to be sufficient for the metrology requirement.

In a similar study, Amigilato *et al.* used CBED to obtain a two dimensional strain map of the active region of a flash memory cell structure with oxide isolation [10]. The electron beam is rastered digitally at a matrix of points in the sample and CBED patterns were recorded at each location. By analyzing the CBED line positions and performing image simulations, the strain tensor components comprising the trace and the shear strain were obtained. Interpolation allowed a strain map for the test structure to be deduced.

3.4. Limitations of CBED

The main advantage of the CBED technique is the high spatial resolution and an accuracy of order 0.01%. However, there are also obvious limitations. The first is that it cannot be applied to a broad area within the sample. This means it is not feasible to generate a strain map of a device in practice. The achievable spatial resolution is constrained by the tilt angle of the sample. Typically, the sample needs to be tilted at a larger angle to minimize the need for taking into account dynamical scattering within the sample. This then allows the use of the simpler kinematic approximation to be applied in simulations of HOLZ lines as discussed in section 3.4. While this may not be an issue in un-patterned samples, it can cause an effective loss of spatial resolution because for a device sample, the beam will be interacting with regions of different strain. For example, the electron beam could pass through both the channel and the source region of a transistor. This can cause the HOLZ lines to lose contrast and become blurred. It has been estimated that the practical spatial resolution of CBED is about 50nm which is not sufficient for current silicon devices [11].

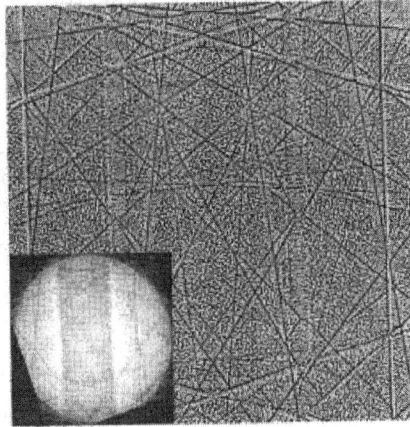

Figure 8: Experimental (inset) and simulated CBED pattern for a STI sample at a location situated at 10μm from the STI structure. Reprinted with permission from M. Kim, J.M. Zuo and G.S. Park, "High-resolution strain measurement in shallow trench isolation structures using dynamic electron diffraction", *Appl. Phys. Lett.,* vol. 84, pp. 2181-2183, Copyright 2004, American Institute of Physics.

Another concern is the relaxation of the sample during sample preparation which can render the measured strain inaccurate. This aspect of CBED had been studied by Clement *et al.* [12]. The above limitations can be overcome by using a different principle called electron holography and this will be the subject of the next chapter.

REFERENCES

[1]   D.K. Schroder, *Semiconductor material and device characterization.* Wiley-Interscience: Hoboken, 2006.

[2]   K. Usuda, T. Mizuno, T. Tezuka, N. Sugiyama, Y. Moriyama, S. Nakahari and S. Takagi, "Strain relaxation of strained-Si layers on SiGe-on-insulator (SGOI) structures after mesa isolation," *Appl. Surf. Sci.,* vol. 224, pp. 113-116, Mar. 2004.

[3]   S.O. Kasap, *Principles of electrical engineering materials and devices.* Mc-Graw Hill: Boston, 2000.

[4]   J.C.H. Spence and J.M. Zuo, *Electron Microdiffraction.* Plenum: New York, 1992.

[5]   K. Usuda, T. Numata, T. Tezuka, N. Sugiyama, Y. Moriyama, S. Nakaharai and S. Takagi, "Strain evaluation for thin strained-Si on SGOI and strained-Si on nothing (SSON) structures using nano-beam electron diffraction," In: *Proceedings of International Electron Device Meeting,* 2003, pp. 138-139.

[6]   K. Usuda, T. Numata, T. Irisawa, N. Hirashita and S. Takagi, "Strain characterization in SOI and strained-Si on SGOI MOSFET channel using nano-beam electron diffraction (NBD)," *Mater. Sci. Eng. B*, vol. 124-125, pp. 143-147, Dec. 2005.

[7]   J.R. Christman, *Fundamentals of solid state physics.* Wiley: New York, 1988.

[8]   K. Kimoto, K. Usami, H. Sakata and M. Tanaka, "Measurement of strain in locally oxidized silicon using convergent-beam electron diffraction," *Jpn, J. Appl. Phys.,* vol. 32, pp. L211-L213, Feb. 1993.

[9]   M. Kim, J.M. Zuo and G.S. Park, "High-resolution strain measurement in shallow trench isolation structures using dynamic electron diffraction," *Appl. Phys. Lett.,* vol. 84, pp. 2181-2183, Mar. 2004.

[10]   A. Amigilato, R. Balboni, G.P. Carnevale, G. Pavia, D. Piccolo, S. Frabboni, A. Benedetti and A.G. Cullis, "Applications of convergent beam electron beam diffraction to two-dimensional strain mapping in silicon devices," *Appl. Phys. Lett.,* vol. 82, pp. 2172-2174, Mar. 2003.

[11]   B. Foran, "Strain measurement by transmission electron microscopy," *Future Fab Intl.* vol. 20, Jan. 2006. [Online] Available: www.future –fab.com. [Accessed Mar. 7 2011].

[12]   L. Clement, R. Pantel, L.F. Tz. Kwakman and J.L. Rouviere, "Strain measurement by convergent-beam electron diffraction: the importance of stress relaxation in lamella preparations," *Appl. Phys. Lett.,* vol. 85, pp. 651-653, Jul. 2004.

CHAPTER 8

Dark-Field Electron Holographic Moire Method

Abstract A strain measurement method based on the quantum interference of coherent electron waves scattered from adjacent strained and unstrained regions of a device is discussed. The method is related to the off axis electron holography technique and is carried out in analytical TEMs equipped with an ultra-stable electron source, Lorentz lens, electron biprism and quantitative HREM software. The hologram obtained by interference of two diffracted beams from a set of lattice planes is analyzed by Fourier transformation to obtain the phase image. The geometric phase analysis used previously for strain mapping in HREM images is performed on the phase image to give the strain with high spatial resolution. This resolution is determined by the hologram fringe spacing. Moreover, all components of the strain tensor can be obtained if more than one diffracted beam is used. The method has been demonstrated for strained silicon device structures and silicon-silicon germanium epitaxial layers.

Keywords: Electron holography, Moire, Fourier transform, Geometric phase analysis, Transmission electron microscope.

1. INTRODUCTION

This is the latest addition to the repertoire of strain characterization techniques for semiconductors. As discussed below, the off axis electron holography technique is based on a wave interference principle for microscopy invented by Dennis Gabor at the University of London in 1948 [1]. Thus, the principle had, in fact, been known for decades. As a microscopy technique, off axis electron holography had been proven useful in various fields such as mapping electrostatic potential within semiconductors [2], mapping dopant composition of semiconductor nanowires [3], dopant distribution in silicon p-n junctions [4] as well as the magnetization in magnetic materials [5]. An excellent review of these applications of off axis electron holography can be found in reference [1]. However, it was only in the last two years that a combined off axis electron holography and optical Moire technique has been demonstrated for extracting strain information in semiconductor materials. Holographic interferometry as it is called for strain measurement was reported in 2008 by Hytch and co-workers at the French national research institute CMES-CNRS in Toulouse [6]. Since its publication, this method has generated much interest and had recently been commercialized.

When compared with the techniques introduced in previous chapters, there are several unique and important advantages for the holographic interferometry method. The optical techniques (spectroscopic ellipsometry, photoreflectance and micro-Raman spectroscopy) can probe a relatively large area of the sample surface. However, their spatial resolution for strain is very limited. The same can be said for the X-ray microdiffraction technique to be discussed in chapter 11. Both the nano beam diffraction and convergent beam electron diffraction methods have high spatial resolution but they can only yield strain information at one very small location of the sample. For strain characterization at say the channel region of a MOSFET, it is desirable to have an electron microscopy technique that combines a much greater field of view with high spatial strain resolution and precision. The holographic interferometry method combines all these attributes and can therefore be considered as a breakthrough development in strain metrology in recent years [6].

2. MEASUREMENT PRINCIPLE

The dark field holographic Moire method involves application of both off axis electron holography and the Moire technique. In the following, we will first present the principle of off axis electron holography and then introduce the holographic Moire principle. This method of measuring strain is an example of using quantum interference of electron waves for an engineering application.

2.1. Off Axis Electron Holography

The interference of coherent electron waves is a striking demonstration of the wave nature of the electron. In physics texts such as the Feynman lectures of physics [7] and in the popular science press, there are lucid

Terence K.S. Wong

accounts of the phenomenon. If two narrow slits are placed between an electron source and a detector screen, then over time, a fringe pattern reminiscent of Young's interference experiment for optical waves will be recorded. This pattern will be seen even if the rate of emission of electrons is reduced to such a level that only one electron passes through the double slit at any instant. The reason is that the interference pattern is due to the wave function associated with the electron. The square of the magnitude of the electron wavefunction after superposition will predict the probability of finding the electron on the detector. This phenomenon can be observed experimentally in an analytical TEM.

Off axis electron holography is an alternative approach to electron microscopy that is based on the coherent or interference of electron waves. In conventional TEM, an electron wave from the source propagates through a thin specimen. As the electrons progress through the sample, it will undergo scattering and the electron wave emerging from the specimen can be written mathematically as [1]:

$$f(\vec{r}) = A(\vec{r})\exp\left(i\varphi(\vec{r})\right) \tag{8.1}$$

$A(r)$ is the amplitude as a function of the position vector $r(x, y)$ in the exit plane and $\phi(r)$ is the phase angle as a function of r. Clearly there are two components to this function. In conventional bright field imaging, the image formed by the lens system of the TEM after the sample is basically an intensity image. This image only makes use of the amplitude component of the electron wavefunction. The phase component of the electron wavefunction is not used and valuable information is in fact lost. This is because from the phase of the electron wave, one can extract (depending on the sample) a wide range of useful sample information such as the electric field strength, magnetic flux density and chemical composition. Samples with poor contrast in the conventional bright field imaging mode can show good resolution and contrast in the phase image and thus the phase image can supplement the conventional bright field image. However, the most significant aspect of phase imaging is that it enables the microscopist to carry out quantitative image analysis.

The goal of off axis electron holography is to generate an electron interference pattern called the hologram from which one can mathematically obtain the amplitude function $A(r)$ and the phase function $\phi(r)$. The hologram is generated by superposing a plane wave which is coherent with respect to the image wave and is derived from the same electron source. Since the plane wave has not passed through the sample, it can be represented as a complex exponential as in equation (8.1) with a constant amplitude and phase angle. The superposition can be performed by adding a component called the electrostatic biprism to the electron microscope column. The biprism was invented by Moellenstedt in 1957 [8]. It creates two coherent virtual sources. After superposition, an interference pattern is formed and the intensity of this pattern is recorded as a hologram by a detector such as a charge coupled device. The hologram intensity can be written as [1]:

$$I_h\left(\vec{r}\right) = 1 + A^2\left(\vec{r}\right) + 2VA\left(\vec{r}\right)\cos\left(2\pi q_c \vec{r} + \varphi\left(\vec{r}\right)\right) \tag{8.2}$$

Here, V is the contrast of the hologram fringes and is determined by several factors related to the microscope. q_c is a wave vector. The significance of the hologram is that the hologram intensity function (8.2) contains both the amplitude function $A(r)$ and the phase function $\phi(r)$ of the image wave. Thus by using mathematical analysis, one can recover both functions and thus deduce more information about the sample than a conventional image. The procedure involved is the Fourier transform (FT) [1]:

$$FT\left(I_h\right) = \delta(q) + FT(A^2) + V.FT\left(A\exp(i\varphi) \otimes \delta(q - q_c)\right) + V.FT\left(A\exp(-i\varphi) \otimes \delta(q + q_c)\right) \tag{8.3}$$

The first and second terms are a Dirac delta function and the FT of the square of the wave amplitude (intensity) of the electron wavefunction respectively. These two terms carry no information about phase and are therefore of no interest in electron holography. The third and fourth terms are conjugate terms (differing in sign only) and are both the convolution of the electron wavefunction with the delta function centered at two spatial frequencies. Either one can be chosen to recover $A(r)$ and $\phi(r)$. In practice, the procedure is basically to select one of these side bands and shift it so that it is centered at $q = 0$ and then take the inverse Fourier transform (IFT) to recover the entire image wavefunction [1]:

$$f(r) = VA(r)\exp(i\varphi(r)) \tag{8.4}$$

Thus, from image analysis of the recorded hologram, one can derive two images for the sample: (i) an amplitude image $A(r)$ and the phase image $\phi(r)$. The latter is important because the phase image can contain information about the electric and magnetic fields within the sample. The phase is related to the electric and magnetic fields because of the following fundamental equation about electron holography [8]:

$$\varphi(x,y) = \sigma V_{proj}(x,y) - 2\pi\frac{e}{b}\Phi_{mag}(x,y) \tag{8.5}$$

In the above equation, σ is called the interaction constant of the TEM and is given by:

$$\sigma = 2\pi\frac{e}{h\upsilon} \tag{8.6}$$

e is the magnitude of the electron charge; h is Planck's constant and υ is the velocity of the electron. V_{proj} in equation (8.6) is called the projected electric potential of the object potential $V_{obj}(x,y)$. $V_{obj}(x,y)$ is the potential distribution within the sample with respect to vacuum. It is also sometimes called the mean inner potential (MIP). Φ_{mag} is the magnetic flux at (x,y).

$$V_{proj}(x,y) = \int V_{obj}(x,y)dz \tag{8.7}$$

Thus, the projected potential is the integral of the object potential with respect to the direction perpendicular to the sample plane. If the sample is magnetic, then the second term in equation (8.5 will also be present. Φ_{mag} caused by the magnetic flux density B_{max} of the sample at an elemental area dA and is given by [1]:

$$\Phi_{mag}(x,y) = \int B_{obj}(x,y,z)dA \tag{8.8}$$

Equation (8.5) shows the reason why off axis electron holography has become an accepted method for characterizing dopant distribution and junction depths within semiconductors. The electrically active dopant ions within the wafer have their own electric potential. This potential perturbs that of the host crystal and thus shows up as contrast in the phase image. Thus, although the source and drain regions of a field effect transistor may show not show up as distinct regions in a conventional bright field image, they will appear as distinct regions in the phase image of the same sample [1]. By the same principle, magnetic domains in recording media can be revealed in the phase image [1].

2.2. Dark-Field Electron Holographic Moire Method

This method is an extension of off-axis electron holography for the purpose of strain mapping. As should be evident from equation (8.5), the phase difference between the image wave and the reference wave does not contain information about strain. In order to obtain this information, Hytch *et al.* showed that one needs to combine off axis electron holography with the optical Moire technique which was discussed in chapter 1. Thus, the dark-field holographic Moire technique is a hybrid approach to strain characterization. The basic principle of this method is illustrated in Fig. **1** [6]. The main difference between this new technique and off axis electron holography is that the reference wave is not a plane wave propagating through the vacuum of the TEM column. Instead, the incident beam passes through a sample consisting of two regions. One region is under strain and the second (the substrate) is unstrained. This is the reason why it resembles the optical Moire technique. In practice, this type of sample can be readily prepared by sample cross sectioning. The electron beam is incident at an oblique angle that corresponds to the characteristic diffraction angle of a set of crystal planes of the sample. As a result, two diffracted beams emerge from the other side of each region of the sample. The electrostatic biprism combines these two diffracted beams as in off axis electron holography and this generates a holographic Moire interference pattern. From this pattern, the phase difference between the electron beams through the strained and unstrained regions of the sample can again be extracted by the procedure described in section 2.1. This phase difference consists of two components. One is dependent on the dynamical elastic scattering and the other is the geometric phase. If the sample has perfectly uniform thickness, the dynamical elastic component is constant and can be considered as an offset term. As will be emphasized in the next section, this is a rather stringent requirement that makes the technique challenging to implement in practice. The second phase component is called geometric phase and this is the component that carries the strain information because it depends on the displacement. The next section describes how the strain map can be derived from the phase image.

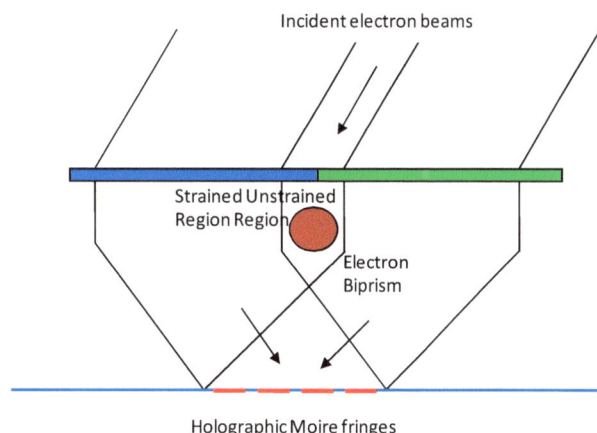

Figure 1: Principle of dark-field holographic Moire method for measuring strain in strained semiconductor devices.

Why is it necessary to introduce electron holography to what is basically the Moire fringe technique? The reason is that the stacked samples needed for the Moire technique is difficult to prepare for strained semiconductor samples. In addition, using only the Moire fringes, it is difficult to obtain narrow fringe spacing and therefore high spatial resolution. Also, by using electron holography, a much larger field of view than that in HREM can be realized when a lower magnification is used for imaging the region of interest. Furthermore, this will not compromise the strain sensitivity and spatial resolution of the strain measurement.

2.3. Geometric Phase Analysis

The geometric phase analysis (GPA) is the mathematical technique used to extract strain information from a dark field hologram. The GPA technique originated from a displacement field analysis procedure developed by Hytch, Snoeck and Kilaas for high resolution transmission electron microscopy (HREM) in 1998 [9]. Initially, the technique was only applied to a very small area of a HREM sample. The following is intended to be a helpful summary of the key concepts in GPA. The full derivation which is quite lengthy can be found in [9] and in a subsequent article [10].

Consider the HREM imaging of a perfect semiconductor crystal. Since this is a periodic structure, the HREM image of this crystal can be written as a Fourier series [9]:

$$I(r) = \sum_g H_g \exp\left[2\pi i g \bullet r\right] \qquad (8.9)$$

Here, $I(r)$ is the intensity of the crystal image at some arbitrary position vector r and g designates the wave vectors of the complex exponential functions with the periodicity of the crystal. H_g represents the corresponding Fourier coefficients which are complex quantities and are written as [9]:

$$H_g = A_g \exp\left[iP_g\right] \tag{8.10}$$

where A_g is the amplitude and P_g is the phase. This phase angle determines the relative positions of all the periodic functions that upon summation will yield the original HREM image. Now suppose the sample is deformed and the image deviates from perfect periodicity. Fourier analysis tells us that the image intensity $I(r)$ of this aperiodic function should be represented by its Fourier transform $I(k)$ as [9]:

$$I(r) = \iint \tilde{I}(k) \exp\{2\pi ik \bullet r\} dk \tag{8.11}$$

By allowing the Fourier coefficients in equation (8.9) to be function of the spatial coordinates, we can write the Fourier transform $I(k)$ as [9]:

$$\tilde{I}(k) = \sum_g H_g(k) \otimes \delta(k - g) \tag{8.12}$$

In this equation, $H_g(k)$ is the Fourier transform of the Fourier coefficient, $H_g(r)$ which is now a function of r and the symbol $\otimes$ denotes the mathematical operation of convolution. In other words, the Fourier transform of the image of the strained crystal is the sum over all Bragg positions of the convolution of $H_g(k)$ and the delta function centered at g. This reverts to the Fourier series representation when the crystal is perfect because $H_g(k)$ becomes a delta function and $I(k)$ has zero value between all Bragg positions. In GPA analysis, we are often interested in the function $H_g(k)$ in the vicinity of a Bragg position (within one Brillouin zone). This can be written in terms of a 'hat' function, $M(k)$ which has unity value within the Brillouin zone being considered and zero value elsewhere.

$$\tilde{H}_g(k) = \tilde{I}(k + g)\tilde{M}(k) \tag{8.13}$$

If we take the IFT of the above function (8.13) for an arbitrary pair of conjugate Bragg positions ($\pm g$) of the deformed crystal, we will generate a 'Bragg filtered' real space image, $B_g(r)$ [9]:

$$B_g(r) = 2A_g(r)\cos\{2\pi g \bullet r + P_g(r)\} \tag{8.14}$$

Note that the factor of two and the cosine is the result of the conjugate symmetry: $H_{-g}(r) = H_g^*(r)$. The Bragg filtered image has an amplitude $A_g(r)$ and a geometric phase component, $P_g(r)$. For the perfect crystal, these quantities will not be functions of position and we can write [9]:

$$B_g(r) = 2A_g \cos\{2\pi g \bullet r + P_g\} \tag{8.15}$$

Starting with this Bragg filtered image of the perfect crystal, we can introduce a displacement field u and make a change of variable to account for the deformation of the crystal [9]:

$$r \to r - u \tag{8.16}$$

The resulting function is:

$$B_g(r) = 2A_g \cos\{2\pi g \bullet r - 2\pi g \bullet u + P_g\} \tag{8.17}$$

However, this must be equal to equation (8.14) above. Therefore, we can compare corresponding terms and conclude that [9]:

$$P_g(r) = -2\pi g \bullet u \tag{8.18}$$

This relationship basically states that a displacement field $u(r)$ has the effect of changing the value of the phase function $P_g(r)$ of the Bragg filtered image. If one can obtain this function, the displacement field can be deduced. In arriving at the relationship (8.18), the arbitrary constant phase P_g in (8.17) has been dropped because it is of no significance. As explained further in [9], there is a complementary way of interpreting the phase function $P_g(r)$ that involves the change in the reciprocal lattice vector. This can be shown to be equivalent to the displacement field given here.

Equation (8.18) can be used to find the components of the two-dimensional displacement field, u. Since there are two independent unknowns, it is necessary to apply (8.18) twice by choosing two non-collinear reciprocal lattice vectors, g_1 and g_2. These can be written as [9]:

$$P_{g1}(r) = -2\pi\{g_{1x}u_x(r) + g_{1y}u_y(r)\}$$
$$P_{g2}(r) = -2\pi\{g_{2x}u_x(r) + g_{2y}u_y(r)\} \tag{8.19}$$

Where $u_x(r)$ and $u_y(r)$ are the components of the displacement $u(r)$ in the x and y directions respectively and $g_{1(2)x}$ and $g_{1(2)y}$ are the components of $g_{1(2)}$ in the k_x and k_y directions where k is the reciprocal lattice vector. The vectors g_1 and g_2 correspond respectively to two vectors a_1 and a_2 in real space [9]. By using a_1 and a_2 instead of g_1 and g_2, the displacement field can be written in vector notation as [9]:

$$u(r) = -\frac{1}{2\pi}\left[P_{g1}(r)a_1 + P_{g2}(r)a_2\right] \tag{8.20}$$

This equation succinctly shows how to obtain the displacement field. First, we choose two non-collinear reciprocal lattice vectors in the Fourier transform of the original image and perform Bragg filtering for each to generate the phase images P_{g1} and P_{g2}. Then we find the real space lattice vectors a_1 and a_2 from g_1 and g_2. The displacement components at any position r in the real space image are given by equation (8.20).

Once the displacement field $u(r)$ is found, the gradient of this field with respect to the x and y directions can be taken and a matrix e can be assembled [9]:

$$e = \begin{pmatrix} \dfrac{\partial u_x}{\partial x} & \dfrac{\partial u_x}{\partial y} \\[2ex] \dfrac{\partial u_y}{\partial x} & \dfrac{\partial u_y}{\partial y} \end{pmatrix} \tag{8.21}$$

The deviatoric strain ε is then found from this matrix by:

$$\varepsilon = \frac{\{e + e^T\}}{2} \tag{8.22}$$

where e^T is the transpose matrix of e. From the matrix e, we can also form an anti-symmetric matrix ω and this gives the rigid body rotation:

$$\omega = \frac{\{e - e^T\}}{2} \tag{8.23}$$

An example of a strain field deduced by this method for silicon germanium is shown in Fig. **2** [10].

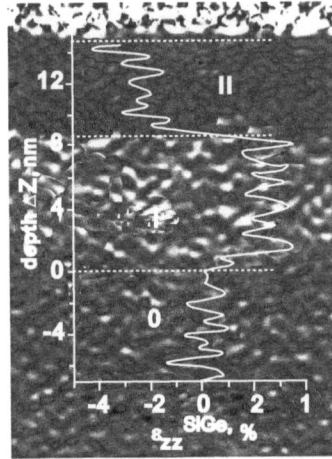

Figure 2: Strain field image ε_{zz} of silicon germanium together with strain profile. Reprinted from *Mater. Sci. Eng. B*, vol. 124-125, N. Cherkashin, M.J. Hytch, E. Snoeck, A. Claverie, J.M. Hartmann and Y. Bogumilowicz, "Quantitative strain and stress measurements in Ge/Si dual channels grown on a $Si_{0.5}Ge_{0.5}$ virtual substrate", pp. 118-122, Copyright 2005, with permission from Elsevier.

2.4. Hologram Analysis Software

The previous section shows that finding the geometric phase image $P_g(r)$ is a pre-requisite to determining the strain. Here we show how this image is obtained. The procedure is available through the Internet as a plug-in software for quantitative electron microscopy software for TEM (www.hremresearch.com). The phase image is found from the Fourier transform of the original image by applying a hat function to one reciprocal lattice vector g. Upon taking the IFT, the following real space image is generated [9]:

$$H_g^{'}(r) = H_g(r)\exp\{2\pi i g \bullet r\}$$ (8.24)

The transform $H_g(r)$ is a complex quantity and so has an amplitude component $A(r)$ and a phase $P_g(r)$. Substituting, we can write:

$$H_g^{'}(r) = A_g(r)\exp\{2\pi i g \bullet r + iP_g(r)\}$$ (8.25)

This means that the phase image $P_g(r)$ is found by:

$$P_g(r) = Arg[H_g^{'}(r)] - 2\pi g \bullet r$$ (8.26)

Where $Arg[H_g^{'}(r)]$ is also called the raw phase image.

3. EXPERIMENT

3.1. Electron Optical Components

In order to successfully apply the dark-field holographic Moire technique, it is necessary to have access to a state of the art TEM with the necessary hardware and software (Fig. **3**). The requirements are similar to those for performing off axis electron holography [11]. First, the TEM should have a coherent field emission source and the power supply of this source must be extremely stable to provide a high degree of fringe contrast in the hologram. A spherical aberration corrector is also required [11]. The second essential hardware component is the electrostatic biprism. An optical biprism is a glass prism where the apex angle is just less than 180 degrees. This allows the light from a single point source to appear to originate from two separate but coherent virtual sources. The electrostatic biprism operates on the same principle but for electron beams [8]. It typically consists of a gold coated quartz wire and this can be biased to around 100V.

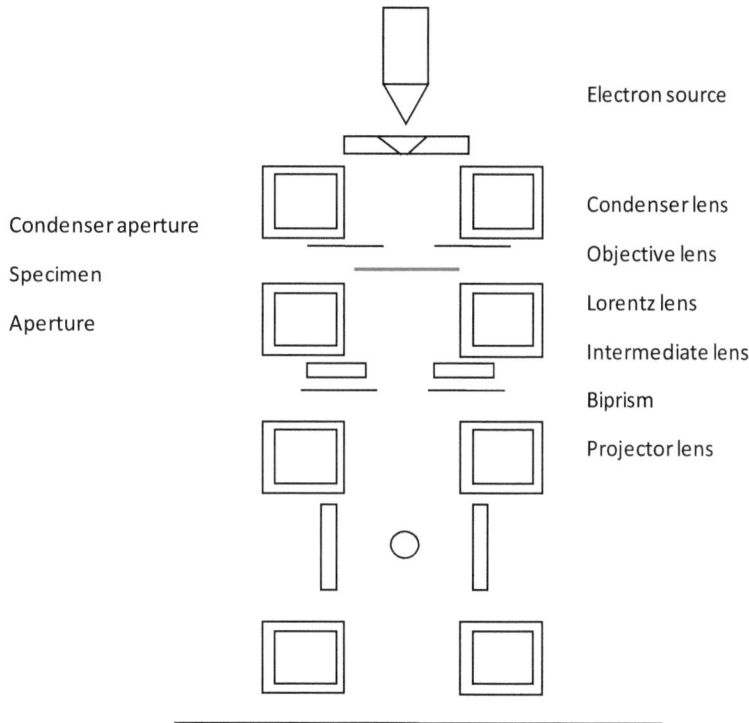

Figure 3: Schematic diagram of TEM electron optical column setup for electron holography.

The third component is called the Lorentz lens [12]. This lens is substituted for the objective lens used for bright field imaging. In other words, during holographic Moire imaging, the objective lens is turned off and the Lorentz lens is used instead to form the holographic image. The Lorentz lens is a low magnification lens. Such a lens is used because for strain mapping, it is frequently desirable to have a larger area or field of view. The Lorentz lens is a mini-lens fitted underneath the lower pole piece of the objective lens. An objective aperture is used to restrict the number of diffracted beams that are allowed to enter the Lorentz lens to form the image.

3.2. Sample Preparation for Holographic Moire Technique

The most demanding aspect of the holographic Moire technique is the sample preparation. In order to obtain strain maps, it is essential to prepare perfectly parallel-sided lamellae samples and there should be no sample bending [12]. In the work by Hytch *et al.* [6], a tripod method was first used to reduce the specimen thickness to several micrometers. These lamellae were then attached by glue to a copper half grid and this was placed vertically within a focused ion beam (FIB) microscope. A gallium (Ga) ion beam was then used to etch the specimen progressively to the final thickness.

4. APPLICATION EXAMPLES

Since the dark field electron holographic Moire technique is quite recent, the number of publications is relatively few. In the following, we will discuss two application example of this technique.

4.1. Strain Mapping in Strained Silicon MOSFET Channels

For demonstration of the holographic Moire technique, Hytch *et al.* used cross sectioned samples taken from a strained silicon dummy transistor array with $Si_{0.8}Ge_{0.2}$ source and drain [6, 13]. The channel length of these dummy transistors is 90nm and the drain current flows in the (110) direction. The SiGe source drain induces longitudinal compressive strains in the channel region. In fully processed devices, this type of strain is used to enhance mobility in p-channel devices.

The holographic Moire images were acquired using a FEI Tecnai F20 ST TEM. Since the longitudinal strain (ε_{xx}) in the (110) direction is of primary interest, the (220) diffracted beam was selected for interference by the biprism to form the dark-field hologram. The resulting hologram has an area of 1µm x 0.25µm. From this hologram, the phase image is calculated and the strain in the (110) direction or ε_{xx} can be extracted by the GPA method. The result is a strain map for ε_{xx}. (Fig. **3** of [6]) According to the GPA theory [9], it is possible to use the holograms generated from two or more diffracted beams to deduce the complete strain tensor of the sample. In this case, the (111) and $\left[\overline{1}11\right]$ were used (Fig. **4** of [6]).

The precision of the strain measurement is determined by using the uniformity of the strain in the Si substrate. Here, there should be no variation in ε_{xx} and the standard deviation of this strain is 0.2%. The accuracy of the strain tensor is checked by using finite element modeling (FEM) method. The agreement with the experimental values is very good. In particular, ε_{xx} varied by about 0.1% in the x direction and ε_{xx} increased monotonically with the distance into the Si substrate. These results show convincingly that strain tensor components can be measured with high precision and accuracy over a large area that is compatible with manufactured strained silicon devices.

4.2. Quantitative Strain Measurements in Epitaxial Silicon Germanium

In this metrological study by Cooper and co-workers at CEA LETI-MINATEC in Grenoble [14], the objectives are to measure strain quantitatively by the dark-field holographic technique and to investigate the accuracy of the strain measured. Two types of samples were grown for this experiment. The first sample consists of a 330nm capping layer of silicon on a 27.5nm layer of $Si_{0.68}Ge_{0.32}$ on a silicon substrate. The second layer consists of a 150nm capping layer of silicon on a stack of alternating Si_xGe_{1-x} (10nm)/Si (30nm) layers with x = 0.45, 0.38, 0.31, 0.20.

Both samples were studied in a state of the art TEM (FEI Titan) with a beam voltage of 200kV. The hologram is generated under dark-field conditions. This means a beam stop is placed at the centre of the condenser lens system to generate a hollow cone of electrons. The crystal planes of interest in this case are the [100] planes in the growth direction. In order to study the strain in the growth direction, the sample had to be tilted to the {400} two beam condition to record the dark field hologram. The use of dark field illumination necessitates the use of a higher beam current and long enough exposure time for recording the electron hologram with sufficient signal to noise ratio. This is why it is essential to have a stable coherent electron source for this type of quantitative electron holography experiment.

For the first sample with a single layer of $Si_{0.68}Ge_{0.32}$, holograms were recorded for three exposure times (4s, 16s, 64s). After obtaining the phase image, strain maps for the [100] planes of the sample were extracted for all three exposure times. It was found that the signal to noise ratio of the strain maps steadily improves with increasing exposure time.

Another important parameter is the spatial resolution of the strain maps. This is determined by the fringe spacing in the recorded hologram. Typically, the spatial resolution of a reconstructed phase image is three times the fringe spacing [14]. Since the sample used always has a strained and unstrained region, the hologram recorded had two fringe spacing. For the sample studied, the spatial resolution is 5nm for the $Si_{0.68}Ge_{0.32}$ layer and 7.5nm for the Si.

The second sample had four layers of $Si_{1-x}Ge_x$ with varying Ge composition separated by Si layers. Thus each layer should have a different amount of strain and in the hologram this will show up as different fringe spacing. From the strain image in the [100] growth direction, the strain in the $Si_{0.8}Ge_{0.2}$, $Si_{0.31}Ge_{0.69}$, $Si_{0.38}Ge_{0.62}$ and $Si_{0.45}Ge_{0.55}$ layers were measured to be 1.5%, 2.0%, 2.4% and 2.8% respectively. These measured strains were compressive and are consistent with finite element simulations. However, the simulations suggest that there should be tensile strain in the thin Si layers between the $Si_{1-x}Ge_x$ layers but this was not observed in the strain map.

REFERENCES

[1]   H. Lichte, P. Formanek, A. Lenk, M. Linck, C. Matzeck, M. Lehmann and P. Simon, "Electron holography: Application to materials questions," *Annu. Rev. Mater. Res.,* vol. 37, pp. 539-588, 2007.

[2]   W.D. Rau, P. Schwander, F.H. Baumann, W. Hoppner and A. Ourmazd, "Two-dimensional mapping of the electrostatic potential in transistors by electron holography," *Phys. Rev. Lett.,* vol. 82, pp. 2614-2617, Mar. 1999.

[3]   M.I. den Hertog, H. Schmid, D. Cooper, J.-L. Rouviere, M.T. Bjork, H. Riel, P. Rivallin, S. Karg and W. Riess, "Mapping active dopants in single silicon nanowires using off-axis electron holography," *Nano. Lett.,* vol. 9, pp. 3837-3834, Nov. 2009.

[4]   D. Cooper, C. Alliot, R. Turche, J-P. Barnes, J-M. Hartman and F. Bertin, "Experimental off-axis electron holography of focused ion beam-prepared Si p-n junctions with different dopant concentrations," *J. Appl. Phys.,* vol. 104, 064513-1-8, Sep. 2008.

[5]   A. Masseboeuf, A. Marty, P. Bayle-Guillemaud, C. Gatel and E. Snoeck, "Quantitative observation of magnetic flux distribution in new magnetic films for future high density recording media," *Nano. Lett.,* vol. 9, pp. 2803-2806, Aug. 2009.

[6]   M. Hytch, F. Houdellier, F. Hue and E. Snoeck, "Nanoscale holographic interferometry for strain measurement in electronic devices," *Nature,* vol. 453, pp. 1086-1089, Jun. 2008.

[7]   R.P. Feynman, R.B. Leighton and M. Sands, *The Feynman lectures on physics vol. III.* Addison Wesley: Reading, 1989.

[8]   E. Volkl, *Introduction to electron holography.* Kluwer Academic/Plenum: New York, 1999.

[9]   M.J. Hytch, E. Snoeck and R. Kilaas, "Quantitative measurement of displacement and strain fields from HREM micrographs," *Ultramicroscopy,* vol. 74, pp. 131-146, Aug. 1998.

[10]   J.L. Rouviere and E. Sarigianidou, "Theoretical discussions on the geometrical phase analysis," *Ultramicroscopy,* vol. 106, pp. 1-17, Dec. 2005.

[11]   N. Cherkashin, M.J. Hytch, E. Snoeck, A. Claverie, J.M. Hartmann and Y. Bogumilowicz, "Quantitative strain and stress measurements in Ge/Si dual channels grown on a $Si_{0.5}Ge_{0.5}$ virtual substrate," *Mater. Sci. Eng. B,* vol. 124-125, pp. 118-122, Dec. 2005.

[12]   D. Cooper, A. Beche, M. Den Hartog, A. Masseboeuf, J.-L. Rouviere, P. Bayle Guillemaud and N. Gambacorti, "Off-axis electron holography for field mapping in the semiconductor industry," *Micros. Analy.,* pp. 5-8, Jul. 2010

[13]   F. Hue, M. Hytch, H. Bender, F. Houdellier and A. Claverie, "Direct mapping of strain in a strained silicon transistor by high-resolution electron microscopy," *Phys. Rev. Lett.,* vol. 100, pp. 156602-1-4, Apr. 2008.

[14]   D. Cooper, J.P. Barnes, J.-M. Hartmann, A. Beche and J.-L. Rouviere, "Dark field electron holography for quantitative strain measurements with nanometer-scale spatial resolution," *Appl. Phys. Lett.,* vol. 95, pp. 0535011-3, Aug. 2009.

PART 4: EMERGING STRAIN METROLOGY

In this final part, we survey three emerging cross-disciplinary strain metrology methods. These methods share a common feature that they are '*in situ*' methods. The strain measurement is carried out within or with the assistance of a microscope or a synchrotron light source. The tip-enhanced Raman spectroscopy method (chapter 9) used the same principle as that described in chapter 5 for thin films. However, the spatial resolution of the Raman spectroscopy is enhanced by using the surface plasmon resonance of a metallic atomic force microscope tip. This near field interaction increases the Raman scattering from a small region beneath the metallic probe but it does not change the far field Raman signal which is still present. This technique is now of great interest to the microelectronics industry and may eventually find application for characterizing strained semiconductor devices. The next method atomic force microscopy digital image correlation (chapter 10) is used mainly by mechanical engineers for MEMS devices. It is an image correlation method for finding the strain distribution in a sample under deformation. The sample is strained and imaged by the atomic force microscope probe at the same time. The last method in this volume is X-ray micro/nanodiffraction. This technique requires a high intensity X-ray beam. As a result, this method requires access to high brilliance synchrotron sources. At present, it is difficult to see how this type of method can be of routine use to the semiconductor industry.

CHAPTER 9

Tip-Enhanced Raman Spectroscopy

Abstract: A strain measurement method based on nanophotonics is described. In this technique, a noble metal probe or nanoparticle film is used to enhance the probability of Raman scattering by using the surface plasmon resonance effect. An incident laser with a wavelength tuned to the surface plasmon resonance excites surface plasmons in a metal nanoparticle. As a result, the electromagnetic field near the probe is enhanced. The region in which Raman scattering is enhanced is in the nanoscale and the spatial resolution of the Raman microprobe is thereby extended. The method requires a scanning probe microscope to be coupled to a confocal Raman microscope and spectrometer. The development of high contrast tip enhanced Raman spectroscopy and tip preparation techniques are discussed.

Keywords: Atomic force microscope, Correlation coefficient, Microtensile test, Polysilicon.

1. INTRODUCTION

In chapter 5, we discussed the principles and applications of conventional micro-Raman spectroscopy for strain characterization of semiconductor samples with known elastic properties. As mentioned, the method came into widespread use after the 1990s to address problems such as the stress fields induced in silicon substrates due to trench isolation structures. Although the micro-Raman technique was initially useful to the semiconductor industry, its limitation has become apparent in recent years. This change is the result of Moore's law on device scaling. As the minimum feature size of devices entered the deep submicron regime during the late 1990s and then the nanoscale in 2004, the spatial resolution limit of micro-Raman spectroscopy had rendered the technique less and less useful for device structures. For the present and future generations of semiconductor devices, the area of interest within a sample that needs to be probed is typically smaller than the micro-Raman probe size. Thus, the technique has to be enhanced to overcome its spatial resolution limitation.

Why is there a minimum probe size limit for the micro-Raman spectrometer? The reason is that in micro-Raman spectroscopy, an incident monochromatic light beam is used to interact with the sample to yield sample information. Since light is an electromagnetic wave, it can undergo diffraction from apertures just like other forms of wave phenomena. When light is diffracted at an aperture, a point source becomes blurred to a spot called an Airy figure with an intensity distribution given by the Bessel function. The Airy figure comprises a circular central Airy spot surrounded by fringes (Fig. **1**). The diameter of the Airy spot is proportional to the ratio $\lambda/n\sin\theta$ where n is the refractive index and θ is one half of the collection angle of the objective lens. The resolution for an optical microscope was formulated by Rayleigh based on the diameter of the Airy spot as [1]:

$$R = 1.22\frac{\lambda}{NA} \qquad\qquad (9.1)$$

where λ is the wavelength of light, $NA = \lambda/n\sin\theta$ is the numerical aperture of the objective lens. The factor of 1.22 came from Rayleigh's definition for the resolution limit. When the first minimum of the Airy figure of one source point is coincident with the centre of the Airy figure of the second point, then the two points can no longer be resolved from one another. Thus, the resolution limit of the microscope in a micro-Raman system is set by the wavelength of the laser light and the numerical objective and is typically around 500 nm [2]. This resolution limit is relatively large in comparison with state-of-the-art semiconductor devices.

There are two approaches to improve the diffraction limited spatial resolution of conventional micro-Raman spectroscopy. The first is to apply near field scanning optical microscopy. This is also called the aperture approach. The second is an apertureless approach called tip enhanced Raman spectroscopy.

2. NEAR FIELD SCANNING OPTICAL MICROSCOPY

The concept behind near field scanning optical microscopy (NSOM) was first postulated by E. Synge in 1928 [2]. The key insight is that the diffraction limit of an optical microscope can be overcome by using an aperture of sub-wavelength diameter and placing the sample very close to this aperture [2]. When this configuration is used, the interaction area is determined by the evanescent waves emanating from the sub-wavelength aperture and this can be in the submicron or nanoscale region. The main component of an NSOM microscope is thus an optical fibre with a sub-wavelength aperture (Fig. **1**). This fibre is used to couple light from the laser and direct the laser light to the area being probed. By scanning the sample relative to the fibre, different areas of the sample can in principle be probed in sequence with sub-wavelength spatial resolution [9, 3].

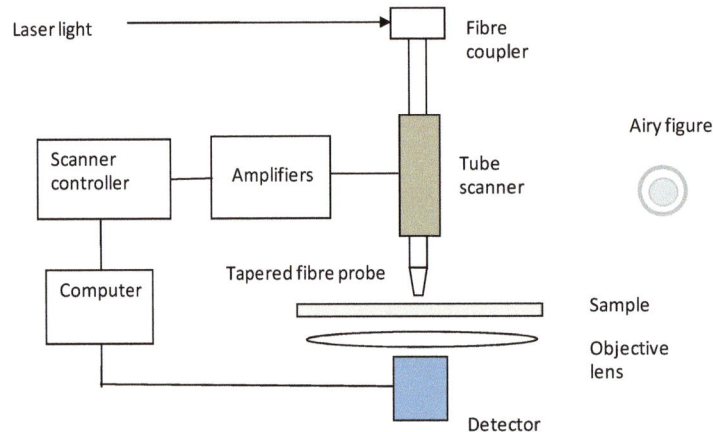

Figure 1: Schematic diagram of a near field scanning optical microscope.

The optical fibre in an NSOM microscope is typically tapered at one end and a reflective metal such as silver is evaporated onto this end in such a way that a very small aperture remains. The reflective metal is used to prevent light from leaking around the aperture. Since the core of the fibre is already small, the aperture will further reduce the photon flux and hence an inherent major problem of the NSOM is that the throughput of the system is very slow. When such a microscope is coupled to a Raman spectroscopy system, the throughput becomes impractical because Raman scattering is intrinsically weak. In one such study by Webster *et al.* on damaged silicon wafers [4], one minute was needed to collect sufficient Raman scattered photons from one location on the sample. Acquiring a complete Raman image from a small area comprising 26 x 21 pixels took nine hours. For these reasons, it is not practical to use NSOM to perform near field scanning Raman spectroscopy. This technique will not be discussed further in this chapter and the interested reader can consult other references [2, 3].

3. SURFACE PLASMON RESONANCE

Before discussing how to improve the throughput of a near field Raman spectroscopy system, it is necessary to make a digression and introduce an important effect called surface plasmon resonance (SPR) [5]. SPR is a unique property of metallic nanostructures made from the noble metals such as gold (Au), silver (Ag) and platinum (Pt). The effect had been used for centuries because this is the method by which the colorful stained glass windows in medieval churches were made. Due to its many modern applications in nanophotonics and as a future interconnect solution in nanoelectronics [6], it is now a burgeoning field of research in its own right. Review articles [7] and complete books [8] had been published. In this section, we will only highlight those aspects of the SPR phenomenon that are especially critical to understanding tip-enhanced Raman spectroscopy.

The term plasmon originally refers to the collective oscillations of the conduction electrons in a bulk metal [5]. These plasmonic waves are similar in concept to the lattice vibrations discussed in chapter 5. According to solid state physics, every metal has a partially occupied energy band called the valence band. Since there are unoccupied energy states in the valence band, the conduction electrons can change state and move freely within the metal. This is why all metals are good conductors and have a resistivity that decreases with decreasing temperature until a minimum is reached. Suppose an electromagnetic field is applied to a metal. Since all the conduction electrons within the metal are charged particles, they will together respond to the action of the electromagnetic field and undergo a collective oscillatory motion. The energy of this collective wave motion as with other types of excitations in solids is quantized. The plasmonic wave with the smallest quantum of energy is called a surface plasmon because the electromagnetic field is confined to the surface of the conductor. For a bulk metal, the plasma frequency ω_p is given by [5]:

$$\omega_p = \sqrt{\frac{Ne^2}{\varepsilon_0 m_e}} \tag{9.2}$$

In this equation, N is the number of electrons per unit volume and m_e is the effective mass of the electron.

Metallic nanostructures exhibiting SPR usually have one of three morphologies: (i) nanoparticles, (ii) nanorods and (iii) nanoshells. Nanoparticles are zero dimensional nanostructures like epitaxially grown quantum dots [9]. All three dimensions of the nanoparticle are in the range of 10-100nm. Both nanorods and nanoshells are one dimensional nanostructures with the radial dimension in the nanoscale. The difference between nanorods and nanoshells is that the nanoshell comprises a core material that is surrounded by a shell made of a different material. The nanorod on the other hand has a homogeneous composition. Many interesting synthetic processes have been developed to fabricate these metallic nanostructures [10].

When light is incident on metallic nanostructures of the noble metals such as Au and Ag, plasmonic effects will be excited. The collective oscillations are confined to the metal dielectric interface or the surface of the nanostructure. At a characteristic frequency called the SPR frequency, large amplitude electromagnetic fields are excited at the surface of the metallic nanostructure. The simplest theoretical model to describe the SPR frequency of a metallic nanostructure is based on solving the Maxwell's equations in electrodynamics. Assuming the nanoparticle is spherical, the Mie theory results in the following expression for the extinction coefficient $k(\lambda)$ [5]:

$$k(\lambda) = \frac{24\pi N_A a^3 \varepsilon_m^{3/2}}{\lambda \ln(10)} \left[\frac{\varepsilon_i}{\left(\varepsilon_r + 2\varepsilon_m\right)^2 + \varepsilon_i^2} \right] \tag{9.3}$$

In the above equation, ε_r and ε_i are the real and imaginary parts of the metal dielectric function respectively; ε_m is the dielectric function of the dielectric medium surrounding the spherical nanoparticle. N_A is the number of nanoparticles per unit area; a is the radius of the nanoparticle and λ is the wavelength. When $\varepsilon_r = -2\varepsilon_m$, a resonance peak will occur in the spectrum of $k(\lambda)$.

4. APERTURELESS APPROACH – TERS

The alternative approach to near field scanning Raman spectroscopy is to use a sharp metallic tip placed just above the sample surface to locally enhance the Raman signal. In early 2000, reports began to appear in the literature about the enhancement of the Raman effect by an atomic force microscope [11]. This effect was found to be useful for the chemical analysis of adsorbed dye molecules on glass substrates [12]. Since then, it has become a widely used technique by the semiconductor industry to study strain in silicon devices with high spatial resolution. Tip enhanced Raman spectroscopy (TERS) was first reported by W.X. Sun and Z.X. Shen at the physics department of the National University of Singapore in 2003 [13]. This microscope was referred as a near-field scanning Raman microscope (NSRM). The experimental setup is based on a conventional micro-Raman spectrometer and a sample scanner (Fig. **2**). The microscope of the Raman spectrometer has an objective lens with high magnification and high NA. The incident laser light is focused by this lens and the Raman scattered photons are also collected by this lens in a backscattering mode. The high NA resulted in a smaller focused probe size and the high magnification facilitates alignment of the metallic tip with respect to the focused laser spot.

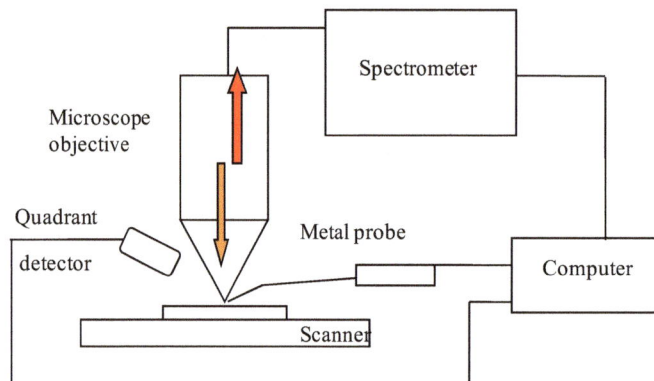

Figure 2: Schematic diagram of the setup of the first NSRM [11].

The sample scanner is similar to those found in every scanning probe microscope. It is basically a piezoelectric actuator capable of translation in the x, y and z directions by nanoscale displacements. The scanner is also equipped with a force sensor to detect the inter-atomic force between the tip and the sample. The scanner is positioned just below the microscope objective lens. During measurements, the sample is placed onto the scanner and the sample is scanned in the x and y directions relative to the metallic tip which is stationary. The signal from the sensor is used in a feedback loop to maintain a constant force and constant separation between the tip and the sample. Thus, in addition to the scanning Raman image, a topographic scanning probe image can be acquired concurrently from the same area of the sample.

Near field scanning Raman spectroscopy is performed by first aligning the metallic tip with respect to the focused laser spot. The scanner is then used to raise the sample towards the tip until the surface is just below the tip for the near field enhancement of the Raman signal. (This process is described further in the next section.) Both the tip and the microscope objective are stationary and are secured to the microscope stage. The scattered photons from the sample are collected by the objective lens. After the Rayleigh scattered photons are filtered, the Raman scattered photons are coupled to the spectrometer for Raman spectrum acquisition. When the Raman spectrum at one location has been recorded, the sample is translated by the scanner to an adjacent location and the process is repeated to yield an area scan of the sample. The acquired Raman spectra can be further analyzed to determine the peak position, peak intensity or integrated intensity. These characteristics can be plotted as a function of spatial position to yield a near field scanning Raman microscopy (NSRM) image. If a Raman peak shift image is plotted, it can be used to quantitatively determine the stress in the scanned area of the sample.

Since both the laser beam of the microscope and the cantilever tip are directed downwards, it is necessary to deflect the beam towards the tip to realize the tip enhancement effect. A simple solution used by Sun and Shen is to simply glue a small mirror at the tip mount and reflect the convergent laser beam to the tip. When the tip is withdrawn by the scanner, the mirror will also be removed from the beam path and conventional micro-Raman spectroscopy can be performed.

5. INITIAL DEVELOPMENT OF TERS

For the initial NSRM demonstration [13], Sun and Shen used a patterned silicon wafer with 380nm wide silicon oxide (SiO_2) stripes at a pitch of 300 nm. Due to the volume change of the silicon upon oxidation, the SiO_2 stripes are slightly higher than the silicon substrate. This provides a sufficient separation between the tip and the silicon substrate during measurement to prevent tip enhancement effects from occurring when the tip is over SiO_2. Raman mapping was performed by placing the cantilevered tip on the sample and exciting it with an argon laser. When the tip is positioned over silicon, the intensity of the Raman peak is greater than that obtained over SiO_2. This is because the Raman spectrum consists of both the near field and the far field components when the tip is over silicon. On the other hand, when the tip is over SiO_2, only the far field component of the spectrum is observed. As a result, the Raman image of this sample shows a contrast due to the near field enhancement of the Raman spectrum by the metal tip.

Another simple way to demonstrate the TERS effect is to withdraw the tip from the surface of the silicon before measuring the Raman spectrum. The Raman spectrum showed a decrease in the peak intensity due to the absence of the near field component. Since it determines the contrast of a tip enhanced Raman image, the amount of enhancement in the Raman spectrum due to the near-field interaction is an important quantity. It is usually called the near-field enhancement factor. The definition used by Sun and Shen for the near-field enhancement factor (EF) is the ratio of the peak intensity of the Raman signal due to the near field interaction (I_{near}) and the intensity due to the far field (I_{far}) only weighted by the respective interaction volumes.

$$EF = \left(\frac{I_{near}}{I_{far}} \right) \left(\frac{V_{far}}{V_{near}} \right) \qquad (9.4)$$

In the above equation, the near-field interaction volume (V_{near}) is determined by the radius of the tip and the interaction depth. The far-field interaction volume (V_{far}) is determined by the diameter of the focused laser spot and the laser penetration depth. Using reasonable estimates, the near-field enhancement factor in this initial demonstration was calculated to be $\sim 10^4$ [13].

Spatial resolution is the other important characteristic of a TERS system. In the setup of Sun and Shen, the spatial resolution was limited by the integration time. During the integration period which was much longer than a typical AFM scan, there was creep and hysteresis of the piezo scanner. As a result, the near-field Raman image was somewhat smeared [13].

6. TOWARDS HIGH CONTRAST TERS

From the above discussion, it is clear that an unavoidable limitation of the TERS technique is the simultaneous presence of a near-field and a far-field contribution to the Raman spectrum. Although the near-field component has higher spatial resolution, it can be difficult to recognize if there is also a very significant far-field component. Thus, improvements had to be found to enhance the contrast in TERS. The term contrast in this context will be defined shortly.

The exclusion of the far-field component in Raman microscopy was first demonstrated by V. Poborchii, T. Tada and T. Kanayama of the National Institute of Advanced Industrial Science and Technology (AIST) in Tsukuba, Japan [14]. These investigators used single crystal silicon (001) substrates to demonstrate subwavelength Raman spectra with the aid of a metal particle. The method is based on the depolarization of the incident light. The key principle of the method is that if the incident monochromatic light is polarized along a specific direction ([110]) of the silicon surface, then the backscattered Raman photons will also be strongly polarized because of the Raman selection rules. If a polarizing component called an analyzer is placed with its axis perpendicular to the polarization of the incident light, then the analyzer can block all the backscattered Raman photons from the substrate and a negligible Raman signal will be detected. This is because in this geometry, the polarization of the backscattered Raman photons is such that they are unable to pass through the analyzer and enter the spectrometer. Now if a light scattering particle is placed onto the silicon surface and illuminated by the laser light, some of the incident polarized light will be scattered by this particle and the degree of polarization will be reduced. As a result, after inelastic scattering in the silicon, some of the backscattered Raman photons will now be able to pass through the analyzer and give rise to a highly localized Raman signal.

In the experiment, Poborchii used 364 nm near UV light to limit the penetration depth of the incident light in silicon to ~ 10nm [14]. The polarized 364nm UV light was focused to form a ~ 500 nm diameter spot at normal incidence. A quartz AFM tip with a 70-100 nm Ag particle on top was then introduced into the laser spot and the Ag particle was positioned onto the silicon (001) surface (Fig. **3**). (A quartz tip was used because it is transparent to the incident light.) In order to further reduce spurious scattering from the AFM cantilever, a drop of glycerol with a refractive index identical to quartz was added to immerse the Ag particle and part of the AFM tip. For the polarizer-analyzer geometry mentioned above, a strong and characteristic first order Raman signal at 521 cm^{-1} was observed for this configuration. However, when the Ag particle is withdrawn, the Raman signal at 521 cm^{-1} became subdued. The spatial resolution of the Raman spectrum obtained by this method was reported to be ~100 nm.

The limitation of this method is that only bare silicon wafers and biaxially stressed silicon epitaxial layers can be measured. This is because they both possess in plane symmetry. For the uniaxially stressed silicon that are used in manufactured devices, the selection rules for Raman scattering are different and the depolarization technique developed by Porborchii *et al.* is not applicable. Hence, the application of this method is mainly for strain mapping in biaxially stressed silicon layers.

Although Porborchii *et al.* mentioned the possibility of plasmonic enhancement of the Raman signal in their paper [14], this was not explored further. The first successful demonstration of high contrast TERS incorporating the depolarization scheme was reported by Lee *et al.* at the University of Akron in 2007 [15]. Their experimental setup was basically similar to that of the AIST group with one important exception that the axis of the Raman microscope objective lens was tilted with respect to the sample axis (Fig. **4**). By using an oblique incidence, one can align the polarization axis of the incident light (514.5 nm) with respect to the AFM tip axis to achieve a maximum plasmonic enhancement of the near field signal under the metalized tip [16]. For the study of the orientation effect of the incident polarized light, the polarizer was rotated with respect to the AFM tip axis. The AFM tip was made from Au or Ag coated silicon nitride cantilever tips.

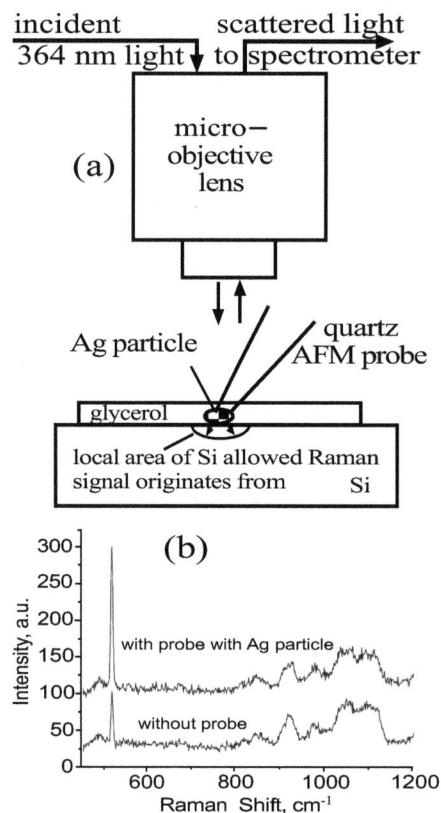

Figure 3: (a) Experiment setup for near field Raman spectroscopy enhanced by a Ag particle attached to a quartz tip; (b) Recorded Raman spectra of silicon with and without Ag particle enhancement. Reprinted from V. Poborchii, T. Tada and T. Kanayama, "Subwavelength-resolution Raman microscopy of Si structures using metal-particle-topped AFM probe", *Jpn. J. Appl. Phys.*, vol. 6, pp. L202-L204, 2005 with permission from Japan Society of Applied Physics.

When the metalized tip was withdrawn from a silicon substrate, the usual Raman signal at 521 cm^{-1} was detected. The intensity of this Raman signal was equal to that of the far field micro-Raman intensity, I_{far}. When the tip was lowered onto the silicon, the intensity of the Raman signal at 521 cm^{-1} increased and the intensity is called I_{total}. This is the intensity observed in the contact state and it consists of several components: (i) far field intensity not shadowed by the tip, (ii) near field enhanced signal because of the tip and (iii) any additional induced signal due to light scattering and reflection from the illuminated portion of the tip. The difference between I_{far} and I_{total} is called I_{near}. This 'near field' intensity is used to define what is called the contrast of TERS and is given by [15]:

$$Contrast = \frac{I_{total}}{I_{far}} - 1 = \frac{I_{near}}{I_{far}} \tag{9.5}$$

Note that this definition does not involve the interaction volume. Using this definition, it was shown that the contrast in TERS can be enhanced by changing the orientation angle between the incident light and the tip axis. This is accomplished through the use of a polarizer. A maximum contrast was achieved when the polarizer is at 70° with respect to the tip axis [15]. The contrast is smallest when the polarizer is at -20° with respect to the tip axis [15]. The difference in contrast between the two extremes is a factor of 7. The contrast can be increased further by placing an analyzer in front of the spectrometer and adjust the analyzer angle relative to the tip axis. When the analyzer angle is set at 90°, the contrast increased to 12 [15].

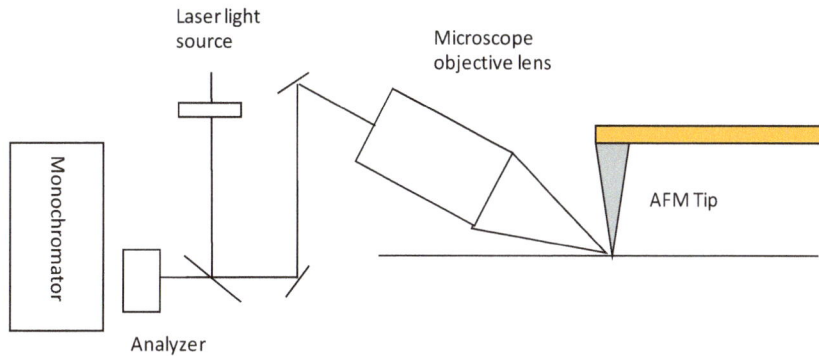

Figure 4: Schematic diagram of the experimental setup for TERS with side illumination optics.

By using optimized polarization and analyzer angles, Lee *et al.* were able to obtain improved contrast in a strained silicon on insulator sample. This sample consisted of 30nm of strained silicon on 100nm of silicon oxide on a silicon substrate. At optimized polarization, the contrast from the strained silicon layer was 350% while that for the unstrained silicon was only 90%. This shows that only contrast increase is confined to the surface region.

N. Hayazawa of the Nanophotonics Laboratory at RIKEN (Saitama, Japan) and researchers at Osaka University used a very similar TERS experimental setup to study strained silicon samples [17]. The experiment was conducted in the reflection mode and the Raman microscope was oriented at an oblique angle with respect to the sample normal. The incident 532 nm light from a YAG laser was first polarized by a half wave plate and a polarizer. It was then focused by an objective lens to deliver a power of 1 mW at the sample. For near field measurements, a metalized silicon cantilever tip from an AFM was positioned over the spot illuminated by the YAG laser. The Raman scattered light was collected by the same long working distance objective lens and was directed into the spectrometer by a dichroic mirror or an unpolarized beam splitter cube [18].

The strained silicon samples used by Hayazawa *et al.* consisted of a silicon substrate with a 2 μm thick graded Si$_x$Ge$_{1-x}$ ($0 < x < 25$ at.%), a 1μm thick buffer layer of Si$_{0.75}$Ge$_{0.25}$ and 30 nm of strained silicon [17]. The near-field Raman spectrum consisted of three peaks: (i) Si-Si peak in the Si$_{0.75}$Ge$_{0.25}$ buffer layer, (ii)

Si-Si peak in strained silicon and (iii) Si-Si peak from the AFM tip. All three peaks are associated with the longitudinal optical (LO) phonon modes of Si-Si bonds. The peak in the buffer layer was by far the most intense while the peak of interest in the strained silicon was the weakest. Thus, the near-field spectrum was not optimized and the unwanted LO phonon modes need to be minimized.

The LO phonon mode at 503 cm^{-1} from Si-Si in $Si_{0.75}Ge_{0.25}$ can be reduced by using a shorter wavelength incident laser. This is because the absorption coefficient of silicon increases rapidly with decreasing wavelength. The light penetration depth in silicon decreases to 5 nm for 351 nm radiation. Thus, by changing from visible light to near UV excitation, one can increase the surface selectivity of the TERS experiment or the intensity of the Si-Si peak in strained silicon relative to that in the Si_xGe_{1-x}. This was observed experimentally when the incident light was reduced from 531 nm to 458 nm and 442 nm.

The LO phonon mode from the unstrained silicon in the AFM tip at 520 cm^{-1} can be eliminated by using a tip made from a dielectric material such as silicon nitride or silicon oxide. The latter can be obtained from a silicon tip by thermal oxidation. By combining the 442 nm emission from a He-Cd laser and a Ag coated silicon nitride tip, the near-field Raman spectrum at a single location was found to be substantially improved. The Si-Si peak from the AFM tip was completely eliminated and the intensity of the peak from $Si_{0.75}Ge_{0.25}$ was reduced relative to the peak of interest. In this way, Hayazawa *et al.* performed the first high contrast, high spatial resolution Raman mapping of local strain fields in epitaxial strained silicon [17]. The image consisted of 128 x 128 pixels. At each pixel, one near-field and one far-field spectrum were acquired. From the near-field spectrum, the Si-Si LO peak was found by Lorentzian fitting and the peak position was plotted as a function of spatial position. After one point was measured, the tip was shifted to the adjacent pixel and the process was repeated to yield the complete TERS image. The near field image showed clear variations in strain in the strained silicon layer. However, when the same area was scanned without the tip, the far-field image showed only uniform contrast and the nanoscale strain variation was not observable.

7. SURFACE ENHANCED RAMAN SPECTROSCOPY

There is another way to perform apertureless near-field Raman microscopy that involves depositing a thin layer of Ag or Au onto the sample. This approach is often called surface enhanced Raman spectroscopy (SERS). Due to the modification of the sample, it is invasive and is therefore less useful than the TERS approach as far as strain metrology is concerned. Hayazwa *et al.* demonstrated SERS measurements for biaxial strained silicon by evaporating a 8-10 nm thick layer of Ag under high vacuum [19]. For these ultra thin layers, the Ag layer was granular and comprised of many contiguous nanoscale grains. The sample was illuminated by blue laser light with wavelength of 488 nm and 9 mW power. The light was focused by an oil immersion, high NA objective lens. The refractive index of the oil was chosen to match that of the lens to result in a smaller focal spot. The main effect of the Ag film is to increase the sensitivity of the Raman measurement to the surface of the strained silicon. This is because localized plasmonic waves are excited at those Ag grains which are located within focused laser spot. The electromagnetic fields from these waves can in turn enhance the Raman scattering from the strained silicon surface. This can be seen by acquiring a Raman spectrum with and without the Ag film from the same strained silicon sample. When there was no Ag film, the Raman spectrum was dominated by the Si-Si Raman peak from the Si-Si bonds in the Si_xGe_{1-x} buffer layer. The Si-Si Raman peak from the strained Si was relatively weak and could only be resolved by performing a Lorentzian fit to the acquired Raman spectrum. The Si-Si Raman peak from the buffer layer was more dominant than the strained silicon peak because of light penetration into the sample. When a Ag film was deposited, the Si-Si peak from the strained silicon in the SERS spectrum was enhanced and was almost as strong as the Si-Si peak from the SiGe buffer layer.

One drawback of the SERS approach is that surface sensitivity is obtained at the price of a reduced scattering efficiency. This refers to a reduction in the Raman counts measured during the integration time and is caused by the presence of the thin Ag layer. This layer also results in a 'bias' in the acquired spectrum because of the scattering by the Ag islands [20, 21].

8. TIP FABRICATION FOR TERS

Since the metalized tip is the most critical part of a TERS measurement system, we provide a brief review of the tip preparation methods for TERS in this section. In the first demonstration of TERS [22], the investigators used an electrochemical etching method to prepare their cantilever metallic tips. The starting material was a drawn polycrystalline tungsten (W) wire with a diameter of about 100 μm. An electrochemical etching kit was used to etch the W wire into a sharp tip. The etchant is either NaOH or KOH and both are aqueous electrolytes. During electrochemical etching, the W wire was connected to the positive terminal of the voltage source and part of it was immersed into the alkaline electrolyte. The positive terminal is called the anode in this case because it is the electrode at which oxidation occurs. The half reaction at the anode is written as:

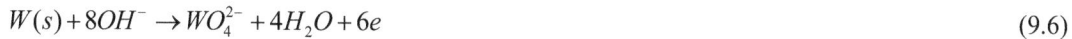

$$W(s) + 8OH^- \rightarrow WO_4^{2-} + 4H_2O + 6e \tag{9.6}$$

This shows that the tungsten metal is oxidized electrochemically by the hydroxide ions in the electrolyte into tungstate ions and water. The reaction will proceed as soon as the applied voltage is greater than 1.43 V. The corresponding reduction reaction at the cathode is:

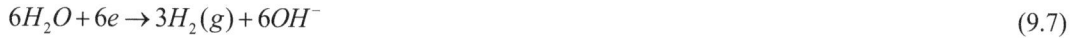

$$6H_2O + 6e \rightarrow 3H_2(g) + 6OH^- \tag{9.7}$$

Here, water is reduced to hydrogen gas and hydroxide ions. These two reactions can be added together to form the overall reaction.

The electrochemical etching reaction occurs at the tungsten electrolyte interface. Due to surface tension effects, a meniscus is formed around the W wire. At the top of the meniscus, the rate of the reaction is considerably reduced because of a concentration gradient effect of the hydroxide ions. The tungstate ions once formed will flow downwards along the length of the wire. These ions form a protective layer around the W wire and impede the reaction between the hydroxide ions and the W. As a result, the rate of reaction further down the immersed portion of the wire is also reduced. Hence the etching reaction is fastest just below the surface of the electrolyte and this causes a bottleneck-like constriction to form in the W wire. The weight of the immersed portion of the W wire will induce stress at this constriction. When the tensile strength is eventually exceeded, the W below the constriction will break off and the etching reaction will be terminated by the controller.

After etching, the W tip was first washed in distilled water to remove the alkali electrolyte. The tips were then coated with Ag by radio frequency (RF) sputtering. Finally, the straight tip was bent into a cantilever by using a home made tool consisting of two sharp blades. The Ag coated tip was placed between the two blades and one blade was slid forward relative to the other to bend the tip into the desired shape.

In subsequent work on TERS, however, investigators tended to use commercial microfabricated Si AFM tips. However, Si tips have two disadvantages. First, as discussed in section 9.6, these tips can contribute an undesired peak in the near-field Raman spectrum that may be close to the near-field peak of interest. Secondly, Si tips are stiffer and may damage the sample being studied. For these reasons, silicon nitride tips were found to be more suitable for TERS measurements. Commercial silicon nitride tips have a lower spring constant and the tip is situated closer to the end of the cantilever. This reduces the shadowing effects of the cantilever and increases the illumination of the tip by the laser light. Before TERS measurements, the silicon nitride tips are metalized by Au or Ag using vacuum deposition. For Au deposition, a chromium (Cr) layer needs to be deposited first to increase adhesion. The thickness and the deposition rate of the noble metal are both critical. If the deposition rate is too high, the cross section profile of the tip may become distorted and this can be determined using transmission electron microscopy. A sufficiently thick layer of Au or Ag should be deposited to cause clustering at the tip. This refers to a roughening of the tip surface when a thicker layer of noble metal is deposited. A tip with clustering can result in a greater enhancement factor.

9. RECENT DEVELOPMENTS IN TERS

Although TERS was only first reported in 2003, the technique had already been commercialized by industry. One company MT-NDT supplies an instrument that combines confocal Raman with AFM and can perform SNOM and TERS. Another interesting recent development is called gap mode TERS. In this implementation of TERS, a metallic probe is again required to generate the near field interaction. However, a scanning tunneling microscope (STM) is used instead of an AFM. This is because the tunneling current of a STM is extremely sensitive to the gap space between the tip and the sample surface. By using tunneling current feedback control, the STM tip can be positioned closer to the surface than is possible with the AFM tip. This enables an even more localized near field interaction. If reflective substrates are used, plasmon resonances within the cavity called cavity modes or gap modes can further enhance the near field Raman scattering. This technique is suitable for samples that do not normally have a strong Raman interaction with light. Howver, thus far, this technique had not been used for nanoscale strain measurements.

REFERENCES

[1]   E.M. Slayte and H.S. Slayter, *Light and electron microscopy.* Cambridge University Press: Cambridge, 1997.

[2]   P.N. Prasad, *Nanophotonics.* John Wiley & Sons, Inc.: New York, 2004.

[3]   Y. De Wilde and P.-A. Lemoine, "Review of NSOM microscopy for materials," In: *Frontiers of Characterization and Metrology for Nanoelectronics,* 2007, pp. 43-51.

[4]   S. Webster, D.N. Batchelder, D.A. Smith, "Submicron resolution measurement of stress in silicon by near-field Raman spectroscopy," *Appl. Phys. Lett.,* vol. 72, pp. 1478-1481, Mar. 1998.

[5]   A.J. Haes, C.L. Haynes, A.D. McFarland, G.C. Schatz, R.P. van Duyne and S. Zou, "Plasmonic materials for surface-enhanced sensing and spectroscopy," *MRS Bull.,* vol. 30, pp. 368-375, May. 2005.

[6]   H.A. Atwater, S. Maier, A. Polman, J.A. Dionne and L. Sweatlock, "The new "p-n" junction: plasmonics enables photonic access to the nanoworld," *MRS. Bull.,* vol. 30, pp. 285-389, May. 2005.

[7]   R. Gordon, "Surface plasmon nanophotonics: A tutorial," *IEEE Nanotech. Mag.,* vol. 2, pp. 13-18, Sep. 2008.

[8]   M.L. Brongersma, *Surface plasmon nanophotonics.* Springer: Berlin/Heidelberg, 2007.

[9]   Y. Xia and N.J. Halas, "Shape-controlled synthesis and surface plasmonic properties of metallic nanostructures," *MRS Bull.,* vol. 30, pp. 368-375, May. 2005.

[10] C.J. Murphy, T.K. Sau, A. Gole and C.J. Orendorff, "Surfactant-directed synthesis and optical properties of one-dimensional plasmonic metallic nanostructures," *MRS Bull.,* vol. 30, pp. 349-355, May. 2005.

[11] R. Stockle, Y.D. Suh, V. Deckert and R. Zenobi, "Nanoscale chemical analysis by tip-enhanced Raman spectroscopy," *Chem. Phys. Lett.,* vol. 318, pp. 131-136, Feb. 2000.

[12] M.S. Anderson, "Locally enhanced Raman spectroscopy with an atomic force microscope," *Appl. Phys. Lett.,* vol. 76, pp. 3130-3132, May. 2000.

[13] W.X. Sun and Z.X. Shen, "Near-field scanning Raman microscopy using apertureless probes," *J. Raman Spectrosc.,* vol. 34, pp. 668-676, Sep. 2003.

[14] V. Poborchii, T. Tada and T. Kanayama, "Subwavelength-resolution Raman microscopy of Si structures using metal-particle-topped AFM probe," *Jpn. J. Appl. Phys.,* vol. 44, pp. L202-L204, Jun. 2005.

[15] N. Lee, R.D. Hartschuh, D. Mehtani, A. Kisliuk, J.F. Maguire, M. Green, M.D. Foster and A.P. Sokolov, "High contrast scanning nano-Raman spectroscopy of silicon," *J. Raman Spectrosc.,* vol. 38, pp. 789-796, Jun. 2007.

[16] O.J.F. Martin and C. Girard, "Controlling and tuning strong optical field gradients at a local probe microscope tip apex," *Appl. Phys. Lett.,* vol. 70, pp. 705-707, Feb. 1997.

[17] N. Hayazawa, M. Motohashi, Y. Saito, H. Ishitobi, A. Ono, T. Ichimura, P. Verma and S. Kawata, "Visualization of localized strain of a crystalline thin layer at the nanoscale by tip-enhanced Raman spectroscopy and microscopy," *J. Raman Spectrosc.,* vol. 38, pp. 684-696, Jun. 2007.

[18] Y. Saito, M. Motohashi, N. Hayazawa, M. Iyoki and S. Kawata, "Nanoscale characterization of strained silicon by tip-enhanced Raman spectroscope in reflection mode," *Appl. Phys. Lett.,* vol. 88, pp. 143109-1-3, Apr. 2006.

[19] N. Hayazawa, M. Notohashi, Y. Saito and S. Kawata, "Highly sensitive strain detection in strained silicon by surface-enhanced Raman spectroscopy," *Appl. Phys. Lett.,* vol. 86, pp. 263114-1-3, Jun. 2005.

[20] N. Hayazawa, Y. Inouye, Z. Sekkat and S. Kawata, "Near-field Raman scattering enhanced by a metalized tip," *Chem. Phys. Lett.,* vol. 335, pp. 369-374, Mar. 2001.

[21] N. Hayazawa, A. Tarun, Y. Inouye and S. Kawata, "Near-field enhanced Raman spectroscopy using side illumination optics," *J. Appl. Phys.,* vol. 92, pp. 6983-6986, Dec. 2002.

[22] W.X. Sun, Z.X. Shen, F.C. Cheong, G.Y. Yu, K.Y. Lim and J.Y. Lin, "Preparation of cantilevered W tips for atomic force microscopy and apertureless near-field scanning optical microscopy," *Rev. Sci. Instrum.* vol. 73, pp. 2942-2947, Aug. 2002.

CHAPTER 10

Atomic Force Microscopy Digital Image Correlation Method

Abstract: An *in situ* method of strain characterization based on image analysis of atomic force microscope images is presented. The sample topographic image is measured before and after deformation using an atomic force microscope with closed loop feedback control of position. By using a mapping function for each point in the image and performing a correlation using the height data, the displacement, strain and strain gradient can be deduced. The digital image correlation algorithm is most suitable for microelectromechanical materials and devices.

Keywords: X-rays, Synchrotron, Beamline, Microdiffraction, Bragg law, Reciprocal space, Electromigration.

1. INTRODUCTION

The atomic force microscope digital image correlation (AFM/DIC) method is an *in situ* method of strain metrology [1]. The test structure which is generally a thin narrow gauge section is attached onto a microtensile test stage and a force is applied to deform the specimen by an actuator. Before and after deformation, an AFM with its probe aligned to the specimen is used to acquire images of an area of the specimen. An example of an AFM topographic image for a biaxial stressed silicon layer on a metamorphic substrate is shown in Fig. **1**. The two AFM images are then compared by an image correlation algorithm by computer. This algorithm basically associates related features on the two scanned images and use a statistical correlation as a criterion to determine the amount of displacement at each point in the surface before deformation. Since the displacement and strain are deduced from image analysis, the method is not restricted to AFM images. It is pointed out that images from optical microscopes and SEMs have also been analyzed by DIC to obtain strain information [2-4].

The AFM/DIC method was developed by Wolfgang Knauss and Ioannis Chasiotis at the Graduate aeronautical laboratory of the California Institute of Technology in the late 1990s to measure the mechanical properties of small mechanical devices and components. Like microelectronics, there had been a steady miniaturization of mechanical components to the micrometer length scale in the past thirty years following the pioneering work of Petersen [5]. These MEMS structures have important applications as sensors and actuators in the automotive, biomedical and wireless communication industries [5]. MEMS can be made by both bulk and surface micromachining techniques [5]. These processing techniques are similar to the photolithography, wet etching and plasma chemical vapor deposition processes that were initially used in microelectronic fabrication. In order to design and measure the performance of MEMS devices and to understand their failure modes for reliability, it is critical to measure the mechanical properties of these devices. These properties include the Young's modulus, Poisson ratio, tensile strength, fracture toughness as well as creep and fatigue behavior [1]. For a number of these properties, a strain measurement on a small MEMS component is vital. However, conventional tensile testing methods are difficult to be applied to such small structures attached to a substrate. As a result, new strain measurement methods are needed for MEMS devices. In an early example of microscale strain measurement, reflective markers are patterned onto a tensile specimen and a laser is used to generate an interference pattern [6]. As the sample is strained, the distance between the markers is increased and this causes a change in the fringe spacing. Thus, by measuring the interference pattern, the elongation and hence unaxial strain can be determined.

Mechanical engineers are also interested in strain measurements on small structures for a more fundamental reason. In the field of experimental mechanics, there is a need to gather experimental data on strain with high spatial resolution to address basic questions such as crack formation and propagation in brittle materials [7]. This data can provide insights for the development of theory in mechanics. As the characteristic length scale of materials reduces to the nanoscale, mechanical properties may become size dependent. Here, again there is a need to understand basic mechanical behavior using *in situ* strain measurements.

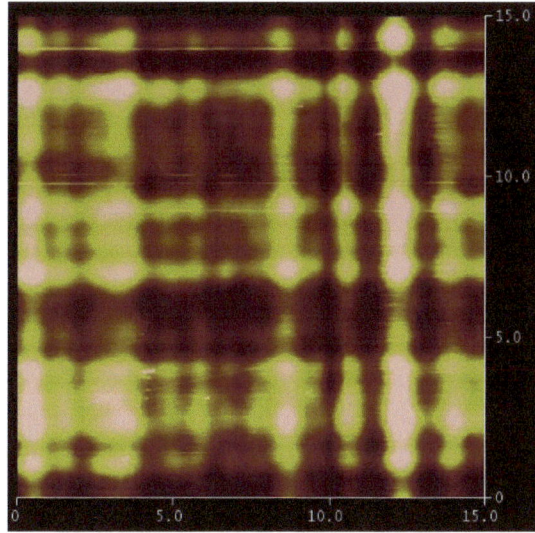

Figure 1: AFM image of strained silicon layer on a silicon germanium metamorphic substrate. The image size is 15μm x 15μm and the color scale shows the height in nm.

2. MEASUREMENT METHOD

2.1. Sutton's DIC Algorithm

The DIC method used to calculate the strain field of a sample from scanned AFM images was based on an early algorithm developed by Sutton and co-workers at the College of Engineering of the University of South Carolina in 1983 [2]. This algorithm was initially applied to the analysis of video images for displacement analysis. The specimen is illuminated by a white light source and images are captured by a stationary video camera. The intensity of light reflected by the specimen is converted into a discrete number of grey levels and this intensity data at a discrete number of locations of the viewed area is stored as a data file for each image. In effect, each digitized image of the specimen is a sampled data set of the reflected light intensity. Before any digital image processing can be performed, this discrete data set must be converted into a continuous form by a mathematical process called interpolation. Interpolation means finding the function that best approximates the true value of an unknown function from sampled values of that unknown function. A bilinear interpolation function was used by Sutton to represent the captured optical images [2]:

$$g(x_1, x_2) = f_d + f_a x_1 + f_b x_2 + f_c x_1 x_2 \qquad (10.1)$$

In this equation, $g(x_1, x_2)$ is the interpolation function for any 2x2 adjacent set of square pixels; f_a, f_b, f_c and f_d are constants to be determined and x_1 and x_2 are distance coordinates. Note that this function is only used to interpolate the intensity data in a very small subset of the image. For another set of four pixels in another part of the image, the constants f_a, f_b, f_c and f_d will be different and have to be calculated again for that subset.

Now suppose the same area of the sample is imaged before and after mechanical deformation and the digitized images are called A and B respectively. It is evident that there must be some mapping relationship between these two images. Specifically, a small subset of pixels from A should be related to another small subset of pixels from B *via* a homogeneous mapping relation. Sutton proposed the use of a two dimensional mapping relation of the form [2]:

$$<x'>_i = <x>_i + <u>_i + \frac{\partial\left(u_i(x)\right)}{\partial x_j} dx_j \quad (i, j = 1,2) \qquad (10.2)$$

x is the position vector of an arbitrary point in the image before deformation and x' is the position vector of this same point of the specimen after deformation. These two positions are assumed to be related linearly by two vector displacements: u_1, u_2 and four displacement gradients: $\partial u_1 / \partial x_1, \partial u_1 / \partial x_2, \partial u_2 / \partial x_1, \partial u_2 / \partial x_2$.

Thus, image correlation involves finding a set of six mapping parameters for a given arbitrary small subset of pixels in image A. In order to do this, the DIC algorithm will search through a larger subset of pixels in the vicinity of the original position (image A) within image B. The search consists of comparing the interpolated intensity data of the chosen subset in image A (equation 10.1) with all other possible subsets of the same size in image B. For each case, the mapping parameters that minimize the sum of the squares of the differences between the two subsets is found and the parameters that give the smallest difference will determine which subset in B is the proper counterpart of the subset in A. A correlation coefficient, C can be defined for this purpose as [2]:

$$C\left(u_i, \frac{\partial u_i}{\partial x_j}\right) = \iint_M \left[A(x) - B(x')\right]^2 dx \quad (i, j = 1, 2) \tag{10.3}$$

Here, $A(x)$ and $B(x')$ represent the interpolated intensity data before deformation (A) and after deformation (B). M designates the subset of pixels used for evaluating the correlation. According to this definition, a minimum value of zero will indicate a perfect correlation between the two subsets. An important point about this early DIC algorithm is that it can only be used for two dimensional deformation analysis and distortions in the out of plane direction cannot be inferred.

2.2. Improved DIC Algorithm

An improvement on Sutton's algorithm was developed by Knauss and Vendroux for scanning probe microscope images [8]. As with video images, each AFM scan results in a digital image that consists of a discrete number of pixels. For each pixel, the topographic height and a number of other sample properties can be measured and associated with that pixel. Thus the AFM image can be considered as a discrete function of two spatial variables x and y. Suppose the AFM image of an arbitrary area of the sample before deformation is applied is represented as $f(x, y)$, where x and y are the Cartesian coordinate axes consistent with the scanned directions. After a tensile stretch is applied, the sample is deformed and the pixels in the initial image $f(x, y)$ would shift to new positions. The AFM image taken after deformation is then represented as $g(x', y')$. A different set of Cartesian axes x' and y' are used to indicate the fact that a deformation has occurred. A one-to-one correspondence obviously exists between the pixels of $f(x, y)$ and $g(x', y')$ for AFM images as well. Suppose the mapping function is χ. The goal of the AFM DIC algorithm is to determine the mathematical expression for χ. In order to find χ, the two discrete image data sets of the sample surface are first converted into a continuous function by using a bi-cubic spline interpolation function. For the deformed surface, the function used to fit the discrete data set is written as [9]:

$$g(x', y') = a_{11} + a_{12}y' + a_{13}y'^2 + a_{14}y'^3 + a_{21}x' + a_{22}x'y' + a_{23}x'y'^2 + a_{14}x'y'^3 + a_{31}x'^2 + a_{32}x'^2 y' + a_{33}x'^2 y'^2 +$$
$$a_{34}x'^2 y'^3 + a_{41}x'^3 + a_{42}x'^3 y' + a_{43}x'^3 y'^2 + a_{44}x'^3 y'^3$$

$$\tag{10.4}$$

In this function, the coefficients a_{ij} are known as the correlation coefficients. A similar bi-cubic spline function can be used to fit the image $f(x, y)$. Note that both $f(x, y)$ and $g(x', y')$ are heights for topographic images. Suppose we consider a point G_p in the initial AFM image. Due to the mapping function, it will end up after deformation at a different spatial location given by $\chi(G_p)$. The mapping function is found by using a least square minimization method to find the best correlation coefficient, C between the two fitted continuous functions $f(x, y)$ and $g(x', y')$ over an area S [9]:

$$C = \frac{\iint_S \left[f(G_P) - g(\chi(G_P))\right]^2 dS}{\iint_S f^2(G_P) dS} \tag{10.5}$$

The correlation coefficient is found by performing a surface integration over the image area and a perfect correlation will yield $C = 0$. The mapping function can also be found by using the cross correlation function [9]:

$$C = 1 - \frac{\iint\limits_{S} f(G_P) g(\chi(G_P)) \, dS}{\left[\iint\limits_{S} f^2(G_P) \, dS \iint\limits_{S} g^2(\chi(G_P)) \, dS \right]^{1/2}} \tag{10.6}$$

The mapping function χ can consist of up to seven parameters. The most basic parameters are the displacement and the first-order displacement gradients. For some materials, the second order displacement gradients are also needed. This is important for MEMS because the second order displacement gradient is in fact a strain gradient. Strain gradients are known to be important for buckling problems in MEMS structures. The last parameter is the out-of-plane deformation or an offset displacement. This parameter is not present in Sutton's algorithm. The DIC method therefore basically involves finding the parameters of the mapping function such that it can best correlate the two given images. From the mapping function, one can obtain the displacement field and calculate the local strain.

The method to find the optimized correlation (minimum C) is illustrated briefly for the simpler case of linear displacement gradients. By substituting the series expansion of $\chi(G_p)$ and taking only up to first order terms, the correlation function can be expressed as a function of the point G_p and a three component vector $P(x) = (u, du/dx, w)$ where w is the out of plane displacement. By expanding the correlation function as a truncated Taylor series and differentiating the result, one can show that at the minimum value of C, the following equation should hold [9]:

$$\nabla\nabla C(P_0)(P - P_0) = -\nabla C(P_0) \tag{10.7}$$

In equation (11.6), P_0 refers to the vector P prior to deformation. This equation can be solved numerically by the Newton-Raphson method. If convergence is achieved, then the value of P will give the u, du/dx and w for the point G_p. Repeated application of this algorithm over other subsets S will yield the components for all points in the image.

3. EXPERIMENT

3.1. Microtensile Test Apparatus

In order to apply the DIC method to MEMS structures, two hardware items are needed. The first is a microtensile test setup and consists of the sample being tested and a force actuator to apply a tension force to the sample. The force applied by the actuator is measured by a force sensor. As elaborated in [10, 11], the sample being tested must be fabricated with the correct shape to yield reliable mechanical data (Fig. **2**). The section being tested is called a gauge section and is the thinnest and narrowest part of the entire test structure. This gauge section is free-standing and is made by depositing the material first as a thin film onto oxidized silicon. After the gauge section is patterned photolithographically and etched, the section is freed at one end by a selective oxide etch using hydrofluoric acid. The other end remains anchored to the silicon substrate.

The free-standing end of the gauge section widens to a large paddle shaped surface for a gripper to be attached. Thus the force is not applied by 'on-chip' actuators. The reason is that MEMS fabricated force actuators are not able to apply sufficient force to deform and break the materials of interest. The area of the paddle must be substantial compared with the gauge section to ensure that the applied force is transmitted uniformly to the gauge section. Also, by having a large area, the stress on the adhesive used to attach the paddle to the gripper can be reduced and this will avoid stress induced relaxation in the adhesive over time. This will ensure that the stress applied will be fully transmitted to the gauge section. For better adhesion, the surface of the paddle is perforated by hundreds of small holes. If the freed gauge section does not have residual stress, then it will lie flat on the substrate and the gripper is simply lowered onto the paddle (Fig. **2**). The gripper is an atomically flat glass surface coated with an ultraviolet (UV) adhesive. When UV light shines through the adhesive, a bond is formed between the gripper and the paddle. If residual stress is present, then a capacitive method will have to be used to pull the paddle flat onto the substrate.

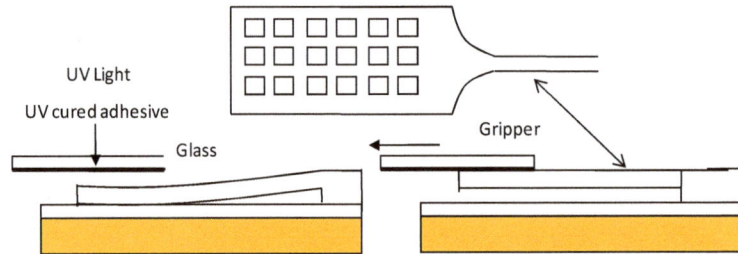

Figure 2: Method of applying the gripper to the microtensile test specimen.

3.2. Atomic Force Microscope for DIC

During strain mapping, the surface of the gauge section is imaged '*in situ*' by an AFM (Fig. **3**). The AFM tip scans over an area in the mid-section of the gauge before a tensile force is applied and after force is applied. As a result of this, the requirements on the AFM are more stringent than that for routine surface imaging. This is due to the principle of the measurement method. In order to obtain reliable and repeatable strain data, it is essential for the AFM to be able to scan over exactly the same area of the sample at least twice in the frame of reference of the AFM tip. This is different from the case for routine surface imaging where the AFM only needs to scan any area of the sample or over some target nanostructure. What is the reason behind this same area requirement? It is because if the AFM tip scans over an area that is displaced from the original area after the deformation is applied by the microtensile setup, the image will contain displacement errors in the entire field and the strain fields derived will be in error. Thus, precautions must be taken to eliminate all sources of spurious displacement errors.

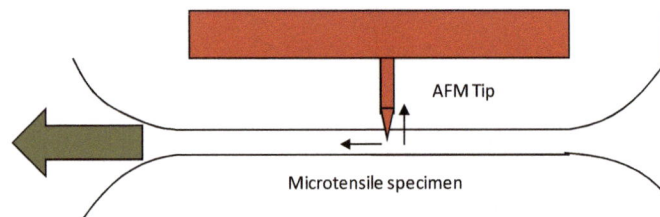

Figure 3: Schematic top view diagram of the thin gauge section of the microtensile test specimen is imaged *in situ* by an AFM.

A main source of displacement error is the hysteresis of the piezoelectric scanner of the AFM tip. The piezoelectric actuator is a crystal that changes its dimensions when a voltage is applied to it. It is the main mechanism at present for scanning the tip of an AFM. However, a limitation of piezoelectric actuators is that when the voltage is reduced back to zero, there is a net displacement due to hysteresis. This hysteresis must be corrected for before the scanner can be used for DIC. Other sources of errors may include thermal drifts and acoustic noise. Thus, the entire apparatus (microtensile stage and AFM) need to be installed in a thermally and acoustically isolated chamber.

The topographic images used for DIC are acquired using the non-contact or tapping mode of the AFM. This is again a precaution to prevent spurious displacements to be induced in the free standing gauge section during measurement. In the tapping mode, the AFM tip is being oscillated at a resonant frequency by an actuator. The change in amplitude or phase of these oscillations is used by the microscope to monitor the force interaction. By minimizing the contact between the tip and the sample, this source of displacement error is eliminated.

4. APPLICATION EXAMPLES

4.1. Polysilicon

Polycrystalline silicon or polysilicon is the most widely used material for fabricating MEMS devices such as sensors and actuators. It can be deposited from silane by plasma enhanced chemical vapor deposition. When the film thickness is above 1 μm, the grains are columnar with a typical grain size in the range of 300 - 600 nm [12]. As a result, the polysilicon film has uniform mechanical properties in the direction perpendicular to the plane of the film. If the area of the film is much greater than the typical grain size, then one can make a further assumption that the in plane mechanical properties are also isotropic. This assumption may however not be valid when the characteristic dimension of the MEMS component is comparable to the grain size such as in cantilever structures. For MEMS applications, it is thus necessary to measure the elastic properties of polysilicon with specimen dimension comparable to the grain size. This enables the determination of the representative volume element (RVE) of the polysilicon [12]. This is the minimum volume for which the assumption that the polysilicon is isotropic is still valid and that the isotropic elastic constants can be applied.

In a recent paper by Cho *et al.* [12], both the Young's modulus and the Poisson's ratio of polysilicon were measured simultaneously and independently by the AFM/DIC method. This is an advancement from prior work where only one property was measured by interferometry and the samples were in form of a free standing membrane [13]. The polysilicon samples were deposited at two independent MEMS facilities. Both sets of independently fabricated polysilicon samples were of the dumb bell shape as described above. They were subjected to tensile testing and *in situ* AFM images were taken during the tests.

For each test sample, five or more load levels were applied and at each load, five AFM images were acquired for performing the DIC. The proper size of the correlation square or the area within which the algorithm will search for the best match between the deformed surface and the original surface was found by iteration. Basically, there must be sufficient pixels in the correlation square to provide enough surface features for the algorithm to generate a true correlation. The optimum correlation square is about 300nm x 300nm. By using this square size, contour maps of the axial and transverse displacements for the gauge section could be generated for each load level. Both displacement plots consist of parallel bands due to the uniform tension. From these plots, the average axial strain is found by performing a linear fit of the axial displacement versus the distance along the axial direction of the specimen. A similar fit was used to find the transverse strain. From the force readings on the load cell, the stress applied to the sample can be determined from the specimen dimensions. Once these three measurements are found, the elastic modulus, Poisson's ratio and the tensile strength can be calculated.

For one set of polysilicon, the average elastic modulus was found to be 164 GPa and is in good agreement with previous measurements by global methods. The elastic modulus of the other set of polysilicon is slightly lower at 155GPa but still consistent with published values. The Poisson's ratio for both sets of polysilicon was nearly the same at 0.219 and 0.224 respectively. Unlike the elastic modulus and Poisson's ratio, the tensile strength of both sets of polysilicon was found to be quite different. The first set has a tensile strength of 3.09GPa while the second set has a tensile strength of only 1.81GPa. This is suggested to be due to the increased surface roughness of the MUMPs polysilicon.

All the mechanical properties measured for the two sets of polysilicon were from 5μm x 15μm AFM images. Although at this scale, the property values are in agreement with an isotropic assumption, polysilicon will cease to be isotropic below a certain domain size. This is RVE mentioned above. It was determined by Cho *et al.* by scanning increasingly small areas within a given 5μm x 15μm area of the sample. It is known that this area size is well described by an isotropic assumption. From the series of AFM images of different scan sizes, axial displacement maps can again be generated and from these, one can extract axial displacement *versus* distance as above. In this case, however, it was found that as the image area decreases, the plot increasingly differ from linearity because of the effect of the microstructure of the polysilicon. Using the average strain obtained from these plots and the far field stress, an effective modulus can be computed for each field and this shows an increase with decreasing domain size. The smallest domain size or RVE consistent with an isotropic assumption for the mechanical properties of polysilicon was 5μm x 10μm.

4.2. Tetrahedral Amorphous Carbon

The second example concerns an emerging material useful for MEMS application called tetrahedral amorphous carbon (ta-C). ta-C is hydrogen free amorphous carbon comprising sp^3 and sp^2 carbon atoms bonded together without any long range order. This type of amorphous carbon only exists in the form of thin films and cannot be obtained in bulk form. The sp^3 carbon atoms have four hybridized (mixed) orbitals for their valence electrons and these are at an angle of 107.8° to each other as in diamond, the crystalline form of carbon. The sp^2 carbon atoms have three hybridized orbitals lying in a common plane and are at 120° to each other. The fourth valence electron occupies a lone p_z orbital perpendicular to the sp^2 orbitals. These sp^2 atoms are also found in graphite and graphene. The overlap of the p_z orbitals enables both graphite and graphene to possess electrical conductivity. Since ta-C consists of both sp³ and sp^2 carbon atoms, it is a material that has high stiffness, hardness and some electrical conductivity. The ratio of sp^3 to sp² carbon atoms will determine the mechanical, electrical as well as optical and thermal properties of the ta-C film. By controlling the deposition conditions, ta-C films with well controlled properties can be deposited and this is why it is a material of great interest to the MEMS research community.

It is important to know the mechanical properties of ta-C films for MEMS applications. However, due to its high Young's modulus, it is difficult to measure its elastic properties because the amount of deformation for a given stress is extremely small. In the paper by Cho *et al.* [14], the AFM/DIC technique was used to measure the Young's modulus, Poisson's ratio. The ability to obtain *in situ* spatial maps of strain with nanoscale resolution was also exploited to understand the fracture of ta-C.

The ta-C films were prepared by the pulsed laser deposition (PLD) of pyrolitic graphite by a KrF excimer laser [14]. The deposition conditions were chosen such that films with a sp^3 content of about 80% were deposited. After PLD, the ta-C films were first annealed in argon gas to relieve the compressive stress in the films. Then the ta-C films were patterned into dumb-bell shaped test structures. The width of the gauge section linking the paddle to the chip was either 10 μm or 50 μm.

Uniform tension was applied to the 10 μm wide gauge sections at incremental load levels. For each load level, six AFM images were acquired for the deformed surface and these were correlated with the corresponding images before deformation. The axial displacement field (in the direction of the applied load) consists of a series of parallel bands in a contour plot as a result of the uniform load (Fig. **4**). The axial displacement data along an axial direction is plotted as a function of distance and the gradient of this plot is equal to the axial strain. For ta-C, this axial strain was found to be in the range of 0 – 1.2%. From the applied loading force and the well cross section dimensions of the gauge section, the stress can be deduced independently. Together, the stress and the strain yield the elastic modulus to be about 759 GPa. This measured modulus is very close to a previous measurement by Brillouin scattering and is much higher than polysilicon.

The Poisson's ratio was deduced by using the transverse displacement field and the axial displacement field. Since ta-C has a very high elastic modulus, the transverse displacement is extremely small and a wider specimen (50 μm) had to be used. The value of ta-C was found to be 0.17 [14]. In addition, the tensile strength of ta-C was measured to be 7.3GPa and the strain at failure was 1.2%. The fracture surface is smooth and consistent with brittle fracture. For the fracture properties of ta-C, wider specimens with a width of 340 μm were fabricated [14]. These specimens contain a central elliptical perforation in the gauge section. During the etching of the perforation, a fine mask was used so that the internal walls of the perforation were very smooth.

4.3. Silica Epoxy Nanocomposites

The third example of the application of the AFM/DIC technique is the mechanical behavior of nanocomposites of epoxy resins and silica nanoparticles [15]. A composite material consists of two or more chemically distinct components blended together. Such materials are usually fabricated when one component alone does not have all the desired properties for an application and a second component is needed to obtain the required set of material properties. For epoxy resins, which are widely used for packaging and sealing of electronic devices, the Young's modulus, creep resistance and glass transition temperature are all adequate. However, these materials are brittle and their fracture toughness (resistance to crack propagation) is not adequate for many applications. Thus, there had been attempts to increase the fracture toughness of epoxy resins by forming a composite with silica particles. After blending the resin with silica particles, the particles can halt the propagation of cracks and enhance the effective fracture toughness.

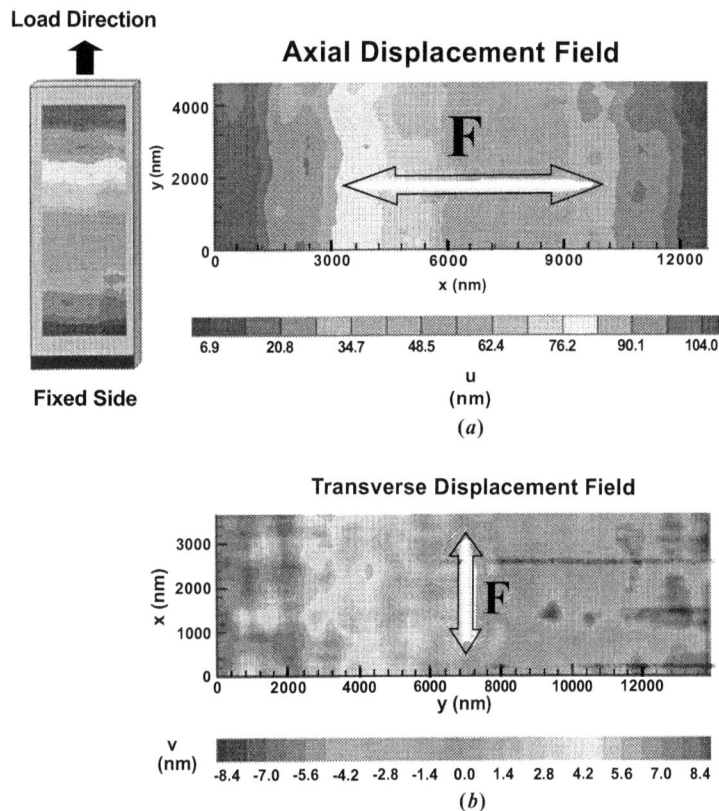

Load Direction

Axial Displacement Field

Transverse Displacement Field

Figure 4: Axial and transverse displacement field of a ta-C specimen subjected to tensile loading. The displacement field was deduced by the AFM/DIC technique [14]. Reprinted with permission 'Young's modulus, Poisson's ratio and failure properties of tetrahedral amorphous diamond-like carbon for MEMS devices', Sungwoo Cho, Ioannis Chasiotis, Thomas A Friedmann and John P Sullivan, J. Micromech. Microeng. 15, 728-735, 2005, IOP Publishing. DOI: 1088/0960-1317/15/4/009.

In a recent work by Chen *et al.* [15], the mechanical properties of an epoxy resin silica nanocomposite had been studied in detail by using the AFM/DIC technique. This work is different from earlier work involving micron scale silica particles. The epoxy resin studied is the thermoset EPON 862 cured by the agent diethyltoluenediamine. Two types of silica nanoparticles were used to fill the EPON 862 matrix at different weight fractions. The first are fumed silica nanoparticles with an average size of 12nm. The second are agglomerates of silica particles with an overall particle size of about 100nm. After dispersing, both composites were then cured by crosslinking using the same curing agent and fabricated as cast plates. From these cast plates, dumb bell shaped specimens with a thin gauge section as mentioned above were carefully machined out. For fracture toughness measurements, it is also necessary to create a notch and a fine initial crack (single etch notch tension specimens).

The AFM/DIC technique allows direct observation of the strain fields in the sample as the specimen is being strained. A uniaxial load was applied to composite samples with 3wt. % and 5 wt.% of 100nm agglomerated silica nanoparticles that resulted in an overall strain of 1%. The strain field in a 3μm x 3μm area was calculated by the DIC algorithm as with the other materials mentioned above. As expected, this was found to be inhomogeneous (Fig. **5**). This strain field was superposed on the corresponding AFM image to show that for both samples, there is a large increase in the local strain in the matrix between the nanoparticles. This is especially the case for the 3wt. % composites where the inter-particle spacing is relatively larger. The local strain in the 3wt. % composites form strain bands with a width approximately equal to the calculated interparticle spacing. This increase in the local strain showed that the inclusion of these silica particles did not make the composite stiffer and thus they did not improve the property of the matrix.

The strain fields of the composites with 5wt. % silica particles are similar to the composites with 3wt. % silica particles. Due to the smaller particle spacing, strain bands were not observed. Nevertheless, the local strain is again much higher than the far field strain and so the elastic modulus of the 5wt. % composites is only slightly higher than the epoxy resin. As for the 12nm silica particles, the resolution of the DIC calculations was not sufficient to give reliable strain maps of the epoxy composites.

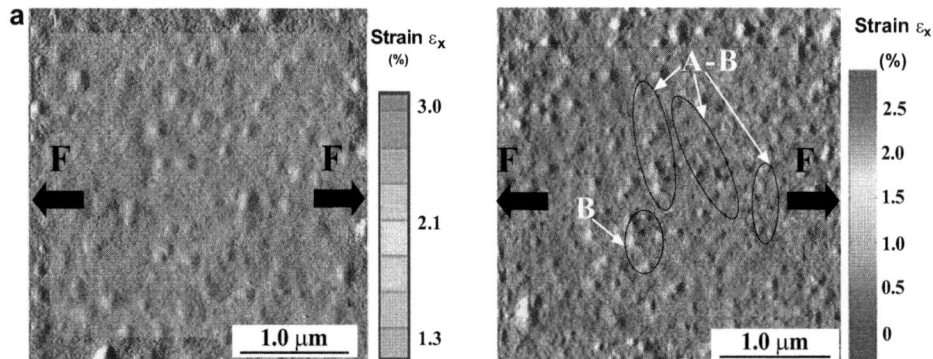

Figure 5: Distribution map of strain ε_x within a 3μm x 3μm region of an epoxy resin nanocomposite with (a) 3 wt.% and (b) 5 wt.% silica subjected to uniaxal tension [15]. Reprinted from *Comp. Sci. Technol.*, vol. 68, Q. Chen, I. Chasiotis, C. Chen and A. Roy, "Nanoscale and effective mechanical behavior and fracture of silica nanocomposites", pp. 3137-3144, Copyright 2005, with permission from Elsevier.

5. SUMMARY

The AFM/DIC method is based on the analysis of digital topographic images before and after deformation. Since its principle rests on the optimization of a correlation function, there are a number of pertinent considerations that should be understood. First, it is crucial to choose (empirically) the optimum subset size *S*. This should neither be too small nor too large. Second, the noise in the AFM signal must be minimized as far as is practically possible. It has been demonstrated that when two identical line scans are made, noise in the signal can fool the algorithm to yield an unphysical displacement. Finally, it is essential for the AFM to operate in close loop so that it is able to scan exactly the same spatial frame before and after the deformation.

REFERENCES

[1]   I. Chasiotis, "Mechanics of thin films and microdevices," *IEEE Trans. Dev. Mater. Rel.* vol. 4, pp. 176-188, Jun. 2004.

[2]   M.A. Sutton, W.J. Wolters, W.H. Peters, W.F. Ranson and S.R. McNeill, "Determination of displacements using an improved digital correlation method," *Image Vision Computing,* vol. 1, pp. 133-139, Aug. 1983.

[3]   W. Tong, "An evaluation of digital image correlation criteria for strain mapping applications," *Strain*, vol. 41, pp. 167-175, Nov. 2005.

[4]   N. Biery, M. de Graff and T.M. Pollock, "A method for measuring microstructural-scale strains using a scanning electron microscope: Applications to g-titanium aluminides," *Metall. Mater. Trans. A*, vol. 34A, pp. 22301-2313, Oct. 2003.

[5]   C. Liu, *Foundations of MEMS.* Pearson: Upper Saddle River, 2006.

[6]   K.J. Hemker and W.N. Sharpe, "Microscale characterization of mechanical properties," Annu. Rev. Mater. Res., vol. 37, pp. 93-126, 2007.

[7]   X. Li, I. Chasiotis, T. Kitamura, "In situ scanning probe microscopy nanomechanical testing," *MRS Bull.,* vol. 35, pp. 361-367, May. 2010.

[8]   G. Vendroux and W.G. Knauss, "Submicron deformation field measurements: Part 2. Improved digital image correlation," *Expt. Mech.,* vol. 38, pp. 86-92, Jun. 1998.

[9]   W.G. Knauss, I. Chasiotis and Y. Huang, "Mechanical measurements at the micron and nanometer scales," *Mech. Mater.,* vol. 35, pp. 217-231, Mar. 2003.

[10] I. Chasiotis, W.G. Knauss, "Microtensile tests with the aid of probe microscopy for the study of MEMS materials", In: *Proceedings of SPIE Vol. 4175*, 2000, pp. 96-103.

[11] I. Chasiotis and W.G. Knauss, "A new microtensile tester for the study of MEMS materials with the aid of atomic force microscopy," *Expt. Mech.,* vol. 42, pp. 51-57, Mar. 2002.

[12] S.W. Cho and I. Chasiotis, "Elastic properties and representative volume element of polycrystalline silicon for MEMS," *Expt. Mech.,* vol. 47, pp. 37-49, Feb. 2007.

[13] S. Cho, J.F. Cardenas-Garcia, I. Chasiotis, "Measurement of nanodisplacements and elastic properties of MEMS *via* the microscopic hole method," *Sen. Actuators A*, vol. 120, pp. 163-171, Apr. 2005.

[14] S. Cho, I. Chasiotis, T.A. Friedmann and J.P. Sullivan, "Young's modulus, Poisson's ratio and failure properties of tetrahedral amorphous diamond-like carbon for MEMS devices," *J. Micromech. Microeng.*, vol. 15, pp. 728-735, Apr. 2005.

[15] Q. Chen, I. Chasiotis, C. Chen and A. Roy, "Nanoscale and effective mechanical behavior and fracture of silica nanocomposites," *Comp. Sci. Technol.*, vol. 68, pp. 3137-3144, Dec. 2008.

CHAPTER 11

Synchrotron X-ray Micro/Nanodiffraction Methods

Abstract: Several X-ray diffraction methods using synchrotron sources are discussed in this chapter. For epitaxial semiconductor samples, high resolution double axis and reciprocal space mapping methods are applicable and measurement time can be reduced by the use of synchrotron radiation. The lattice strain can be deduced directly from the measured inter-planar spacing using Bragg's law without modeling. Polycrystalline samples such as metal films are measured by the $\sin^2\psi$ and $\sin^2\varphi$ methods. Interconnects with widths of the order of one micron can be mapped at this spatial resolution by a scanning X-ray microdiffraction technique. This method does not involve sample rotation by a goniometer. The strain tensor is deduced from an analysis of the von Laue diffraction image and comparison with a structural model of the sample. A high flux density polychromatic X-ray beam is essential for this method.

Keywords: Electron backscattering, Photovoltaics, Lithium ion battery, Resolution, Throughput.

1. INTROUDCTION

X-ray metrology is an extremely powerful set of techniques that is widely used in microelectronics for both device and interconnect characterization [1]. In addition to diffraction methods discussed in this chapter, there are: X-ray fluorescence (XRF) [2], X-ray reflectivity (XRR) [3], near edge X-ray absorption spectroscopy (NEXAS), extended X-ray absorption fine structure (EXAFS) [4] and X-ray scattering (XAS). XAS is fundamentally similar to X-ray diffraction (XRD) [5]. However, it is typically used for non-crystalline solids and colloidal solutions [6]. On the other hand, XRD is applied for single crystals and polycrystalline materials.

X-rays are electromagnetic radiation with a free space wavelength from 4 nm to 0.1 nm [1]. While soft X-rays with wavelength around 4nm could be used for nanolithography, it is hard X-rays with wavelength of order 0.1 nm that are used for structural characterization. This is because their wavelength is comparable to the inter-atomic spacing and thus can resolve crystal structure by diffraction methods. Since X-rays are high energy radiation, they have much greater penetration depth and thus can probe both surface layers and the substrate. By contrast, all scanning probe methods (*e.g.* AFM) can only probe a surface. However, the fact that X-ray photons have higher energy also renders a weaker interaction with matter. In practice, this means it can take considerable time to acquire XRD data with sufficient signal to noise ratio. One solution to this limitation of X-ray methods is to use a bright light source called synchrotrons to perform diffraction experiments. Another advantage of synchrotron sources is that the wavelength of the X-rays can be tuned or selected. This aspect of synchrotrons is vital for other X-ray techniques such as XAS. Some synchrotron beamlines today are also equipped with X-ray optical components that can focus X-rays to a submicron spot size. This capability allows spatial mapping of the properties of a sample and is the reason why the techniques to be discussed in this chapter require use of synchrotron radiation.

In this chapter, we discuss the use of XRD to measure strain in semiconductor technology. XRD can measure strain data because strain is basically a normalized change in the separation between the atomic planes of a crystal. By measuring the inter-planar spacing for the unstressed and stressed regions, the strain can be deduced directly by calculation. The method is fundamentally the same as the NBD technique discussed in chapter 7. Note that unlike the optical techniques of part 2, no modeling of the measured data is needed. Thus, the strain tensor components deduced are not dependent on the assumptions of the models used. This is another major advantage of X-ray methods.

The XRD techniques of this chapter are divided into two parts. First, we will discuss high resolution XRD methods that are widely used for measuring strained epitaxial layers such as strained Si and silicon germanium on single crystal substrates. The methods include double axis diffractometry and reciprocal space mapping (RSM). Next, we will discuss several methods that are applicable to polycrystalline metal layers that are important for interconnect applications. The methods surveyed include: $\sin^2\psi$ method, $\sin^2\varphi$

method and the scanning X-ray microdiffraction method. All these methods are best carried out at third generation synchrotron sources or at a fourth generation free electron laser facility.

2. PRINCIPLE OF XRD

Diffraction is a characteristic phenomenon of all wave motion, including electromagnetic waves [1]. When an X-ray photon arrives at a crystal, the electromagnetic field of the X-ray will interact with the electrons of the atoms inside the crystal. The electrons will gain energy and are excited to higher energy levels. As they de-excite, the excess energy is re-emitted as radiation in all directions. This is termed scattered radiation. The X-ray wave amplitude at any point is the sum or superposition of the scattered radiation from the multitude of electrons within the sample. This superposition must take into account of the phase difference between different point sources and for a crystal result in directions in which there is a high intensity of X-rays and directions in which there is no intensity. For those directions with a strong diffracted beam, the scattered X-rays are in phase and thus can reinforce each other.

2.1. Kinematic Theory of X-ray Diffraction

There are two equivalent ways to formulate the physical principle behind X-ray diffraction. These are the kinematic theory of X-ray diffraction. The basic assumptions in this theory are that there is no energy loss or change in wavelength when the X-ray interacts with the electrons and each X-ray photon is scattered only once. The interaction is weak and so the scattered intensity is also weak. The first formulation is called the Bragg equation named after Lawrence Braggg and William Bragg who discovered this principle in the early twentieth century. Consider the case of X-ray scattering from the atoms at adjacent planes of a perfect crystal (Fig. **1**). The inter-planar distance is d and the wavelength of the monochromatic X-rays is λ. In those directions where the path difference between scattered waves from adjacent layers is an integer multiple of λ, constructive interference will occur and a diffracted beam can be observed. In all other directions, this condition cannot be fulfilled and destructive interference occurs instead. As a result, a diffracted beam is only observed in those directions that satisfy [1]:

$$n\lambda = 2d\sin\theta \tag{11.1}$$

where n is an integer and θ is the diffraction angle. This is by convention defined to be the angle between the beam and the plane of the crystal. It is also the angle between the incident beam and the surface of the crystal. This is why the diffracted beam is also sometimes referred to as the 'reflected' beam. The angle between the incident beam and the diffracted beam is 2θ.

The alternative statement of the diffraction equation is based on a reciprocal vector equation first put forward by von Laue in Germany. The derivation of the following equation can be found in the textbook by Cullity [7]:

$$\frac{S}{\lambda} - \frac{S_0}{\lambda} = H_{hkl} \tag{11.2}$$

Here, S and S_0 are unit vectors in the direction of the diffracted and incident beam respectively. The magnitude of the two vectors on the left hand side has the unit of inverse length and thus it is called a reciprocal vector. The reciprocal lattice vector, H_{hkl} is perpendicular to the (hkl) plane of the crystal where hkl are the Miller indices and the magnitude of H_{hkl} is equal to the reciprocal of the inter-planar spacing d (Fig. **1**).

In the von Laue formulation, diffraction in a certain direction can occur if the vector difference between the outgoing and incident reciprocal vectors is equal to a reciprocal lattice vector. A graphical representation of this is called the Ewald sphere. The radius of this sphere is equal to S_0/λ. By translating the reciprocal lattice vectors of the crystal so that one end is coincident with S_0/λ, the diffraction directions can be found graphically. If the other end of the reciprocal lattice vector terminates at a point of the Ewald sphere, that point will define the direction of the diffracted beam. This principle will be used in the scanning X-ray microdiffraction technique to be discussed in the latter part of this chapter.

The kinematic theory can be readily applied to polycrystalline and powder samples in materials analysis. However, for high quality epitaxial layers, it is not adequate and that is because one or more of the assumptions in the kinematic theory (such as no multiple scattering) breaks down. For these samples, it may be become necessary to use the dynamical theory of X-ray diffraction. This theory can be found in a number of references [1, 5].

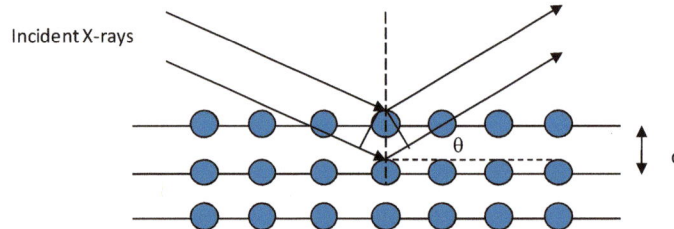

Figure 1: X-ray diffraction from a set of crystallographic planes in a single crystal.

3. SYNCHROTRON RADIATION SOURCES

The strain characterization techniques to be discussed below are carried out in synchrotron radiation (SR) facilities. Although a SR source is not mandatory for high resolution XRD of epitaxial samples, they are essential to microdiffractiion experiments of polycrystalline metal samples. Currently, these are about 40 such large scale research facilities worldwide.

SR refers to the electromagnetic radiation emitted when a relativistic electron is accelerated by a bending magnet or a wiggler device. In current SR facilities, the relativistic electrons are accelerated to high energy by a linear accelerator and then it is injected into a storage ring in which the bending magnets or wigglers are installed. The acceleration due to the change of direction results in the emission of a broadband electromagnetic radiation called SR. The wavelengths range from hard X-rays through ultraviolet to the infrared. This radiation is emitted in a cone with an axis that is tangential to the electron beam path. Thus, it resembles a vehicle headlight when the vehicle turns around a corner. The critical parameter of a SR source is the brilliance which is defined as the number of photons per second per mm^2 per $mrad^2$ per 0.1% bandwidth [8]. A high brilliance source results in high flux and thus greatly reduces the experimentation time.

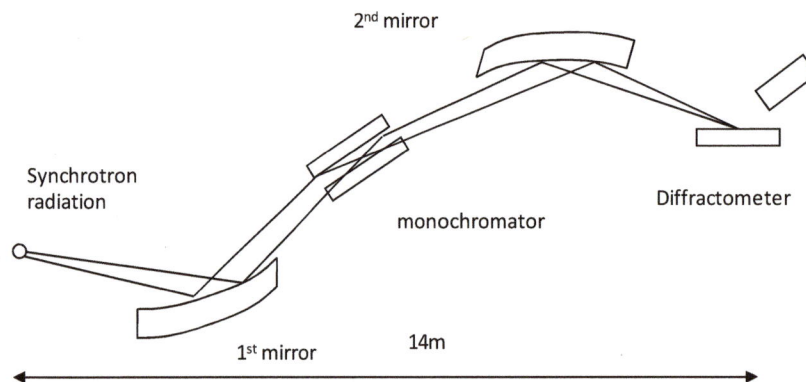

Figure 2: Schematic diagram of the XDD beamline of SSLS for high resolution XRD.

The SR radiation emitted from the storage ring travels down a tubular structure called a beamline to the experimental end station where the experiment is performed. A SR facility typically has multiple beam lines and each is dedicated to a particular region of the electromagnetic spectrum. Inside the beamline are optical components such as X-ray mirrors to condition the beam before it reaches the sample. The equipment and sample are usually housed in a small structure called the hutch. Fig. **2** shows a high resolution X-ray diffractometer housed inside the hutch at the end of the XDD beamline of the Singapore Synchrotron Light Source (SSLS). This instrument is used to obtain the diffraction results shown in section 4.1 of this chapter.

4. DOUBLE AXIS DIFFRACTOMETRY

Double axis diffractometry is a widely used method to characterize epitaxial layers (Fig. **3**) [5]. If the epitaxial layer is not lattice matched to its substrate as in silicon germanium, the lattice spacing will be different for the epilayer and its substrate in the growth direction. The name double axis diffractometry comes from the fact that a highly monochromatic X-ray beam is needed for this measurement and this is obtained by directing the X-rays emitted from a laboratory source (such as Cu K_a source) at a high quality single crystal monochromator. The first axis is the axis of rotation of this monochromator. It allows a highly conditioned reflected beam with small divergence to be selected for incidence on the sample. The sample is rotated independently about a second axis which is perpendicular to the plane of incidence. This plane is defined by the incident beam, reflected beam and the surface normal. Before a double axis diffraction experiment is performed, the source and the detector must be scanned for Bragg peaks. The detector is then positioned at one of these peaks and the sample is rocked about the second axis to generate the so-called rocking curve. During this rotation, the substrate and the epilayer successively satisfy the Bragg condition and the detector will show an increased intensity. This is because the substrate and epilayer have slightly different reciprocal lattice vectors perpendicular to the plane of the substrate. As the sample is rotated, they will successively intersect the Ewald sphere and give rise to diffraction.

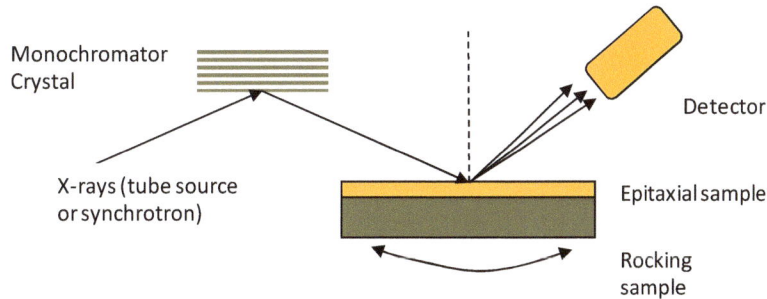

Figure 3: Diffractometer setup for a rocking curve or double axis diffractometry experiment.

4.1. Application Example

The samples described in this section were those mentioned before in chapter 3 on spectroscopic ellipsometry [9]. All three samples S15, S20 and S25 were biaxial stressed silicon on a silicon germanium virtual substrate. HR-XRD rocking curves were acquired for the samples for the (004) Bragg peak (Fig. **4**). For each curve, the large centre peak ($\theta = 0$) corresponds to the silicon substrate. The side peaks located at $+10^3$ to $+2\mathrm{x}10^3$ arc sec are due to the strained silicon layers and those at about -10^3 arc sec are due to the silicon germanium cap layer. In between the silicon germanium cap layer peaks and the silicon substrate peak is the graded silicon buffer layer. The intensity of the strained Si layer is much weaker than the substrate and it occurs at a higher angle than the substrate. This is because the relaxed SiGe cap layer has a larger lattice constant in plane than the Si. The Si is therefore tensile strained. Since the unit cell volume is invariant, the out of plane lattice constant must decrease. According to Bragg's law, the diffraction from the (004) planes of the strained Si must occur at a higher angle. From the SiGe cap layer peak, the out of plane lattice parameter can also be calculated using the Bragg equation. Assuming the SiGe layer is relaxed, the Ge composition can be calculated using (11.3):

$$a_{Si_{1-x}Ge_x} = 5.431 + 0.2x + 0.027x^2 \qquad\qquad (11.3)$$

The Ge composition *x* obtained from HR-XRD for S15 and S20 are listed in the last column of Table **1**. By using the out of plane lattice spacing for the strained Si layer, one can determine the normal strain component in this direction. Using the Poisson ratio, one can deduce the in plane tensile strain as shown in the third column. The strained silicon layer thickness and surface roughness in the first two columns of Table **1** were found from X-ray reflectivity measurements and data fitting [9].

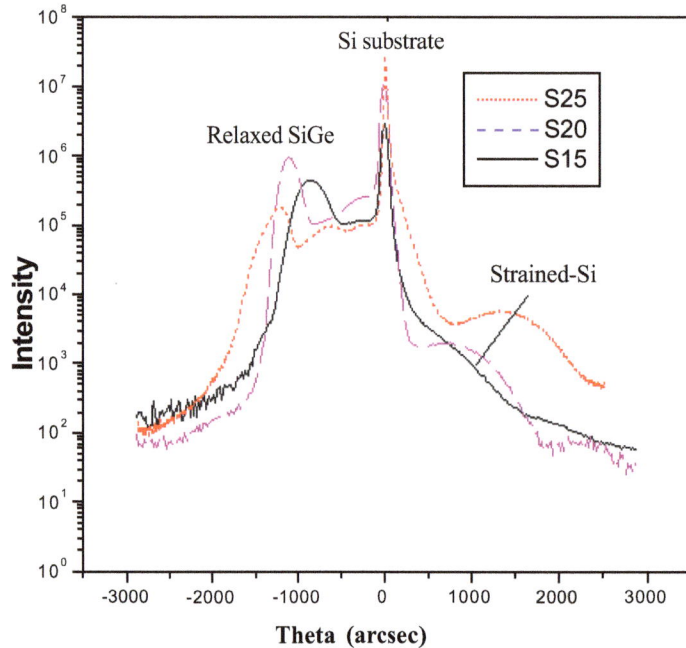

Figure 4: High resolution x ray diffraction rocking curve for ε-Si samples about the (004) Bragg peak for Si. Reprinted with permission 'Characterization of biaxial stressed silicon by spectroscopic ellipsometry and synchrotron X-ray scattering', T.K.S. Wong, Y. Gong, P. Yang, C.M. Ng, Semicond. Sci, and Technol. 22, 1232-1239, November 2007, IOP Publishing. DOI: 10.1088/0268-1242/22/11/009.

Table 1: Strained-Si Sample Parameters Determined from X-ray Reflectivity and High Resolution X-ray Diffraction

Sample	Thickness (nm)	Roughness (nm)	In-plane strain	Ge composition (at.%)
S15	22.5	1.23	0.58	16
S20	21.6	1.44	0.76	21
S25	24.6	>4	1.1	22

5. TRIPLE AXIS DIFFRACTOMETRY AND RECIPROCAL SPACE MAPPING

The experimental arrangement for triple axis diffractometry involves placing an analyzer crystal in front of the detector in the double axis diffractometry set-up. As a result, there is a monochromator crystal conditioning the source X-rays and this can be rotated about the first axis. The sample can be 'rocked' about the second axis and the analyzer crystal can be rotated about the third axis (Fig. **5**) [5]. The reason for adding the analyzer crystal is that like the monochromator, it can be used to precisely select (by diffraction) a very narrow angular range of X-rays of a few arcseconds. This size of acceptance angle cannot be realized by using mechanical slits alone. If one can determine the wave vectors of the reflected X-ray and the incident X-ray very precisely, then the scattering vector can be calculated. By varying the analyzer crystal, the intensity recorded by the detector can be plotted as a function of scattering vector in reciprocal space. The resulting pattern is called a reciprocal space map (RSM) and this is very useful in evaluating the quality of an epitaxial sample.

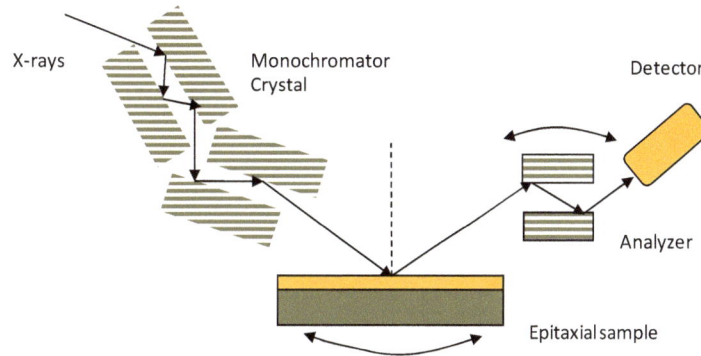

Figure 5: Diffractometer setup for triple axis diffractomery and reciprocal space mapping.

Usually, before a RSM is taken, the sample and the detector are moved to angular positions that correspond a particular Bragg peak. Thereafter, the analyzer crystal is rotated to probe a small volume of reciprocal space around this peak. In the ideal case of a perfect bulk crystal, only a single point should be seen in the RSM. However, in practice, the RSM for a bulk and an epitaxial sample have additional features and these are explained briefly below. The additional features of the RSM are due to two causes. First, there are limitations in the quality of the monochromator and analyzer crystals. Since they are not able to give X-ray beams with infinitesimally small acceptance angles, a good quality sample crystal can result in streak patterns in the RSM. In the simplest case of a bulk crystal, there can be a monochromator streak and an analyzer streak. The second source of additional features is the defects or strain in the sample. This is commonly observed in heteroepitaxial layers as discussed below. Both defects and strain changes the local lattice parameters. As a result, an ideal Bragg peak in the RSM will turn into a diffuse scattering pattern surrounding this point. The extent of this diffuse pattern can be used to gauge the level of defects in an epitaxial sample.

5.1. Application Examples

The strain within a nanoscale strained silicon transistor with silicon germanium source drain stressors was studied by synchrotron nanodiffraction by Cai at the Advanced Photon Source of the Argonne National Laboratory, US in 2007 [10]. In order to overcome the difficulty caused by the extremely small scattering volume of the strained channel region, a high photon flux density beamline and zone plate X-ray focusing optics was used. The flux density of the X-ray beam was 10^4-10^5 photons $s^{-1}nm^{-2}$ for each 0.01% of source bandwidth. By using both a high flux density and small probe size, the diffraction from the channel region can be distinguished from the bulk silicon substrate. Cai mapped both the strained silicon and the silicon germanium stressor by RSM around the (004), (115) and (-115) diffraction peaks. These maps showed that there is lattice bending on both sides of the silicon/silicon germanium interface and strains as a function of lattice curvature were extracted.

In addition to strain measurements, RSM is sometimes useful for epitaxial wafer diagnostic purposes. An example of this is illustrated in Fig. **6** [9]. Here, two of the samples S15 and S25 mentioned in section 4.1 were characterized by triple axis diffractometry. The sample S25 has significantly more Ge than the other sample. RSM was performed for both sample S25 and S15 using the (004) Bragg peak. Fig. **6a** shows the RSM for sample S15 for the (004) diffraction. In this plot, the horizontal axis is the offset angle corresponding to each rocking scan which is displayed along the vertical axis. It is evident that the ε-Si and Si substrate crystallographic planes are almost exactly aligned and there is only a very small tilt of ~0.02^0 between the two. By contrast in Fig. **6b**, there is a larger misalignment between the peaks for the ε-Si and the Si substrate. The tilt angle of the epilayer as determined from this plot is 0.13^0. On the other hand, the ε-Si layer is aligned with the relaxed SiGe buffer. The RSM data thus strongly suggests that the substrate used for the growth of sample S25 is vicinal. As described in [9], this interpretation is supported by X-ray reflectivity and atomic force microscopy measurements. This measurement explained why the quality of the wafer S25 is inferior to wafer S15. Similar RSM measurements on silicon germanium virtual substrate have also been reported by Zhang *et al.* in [11].

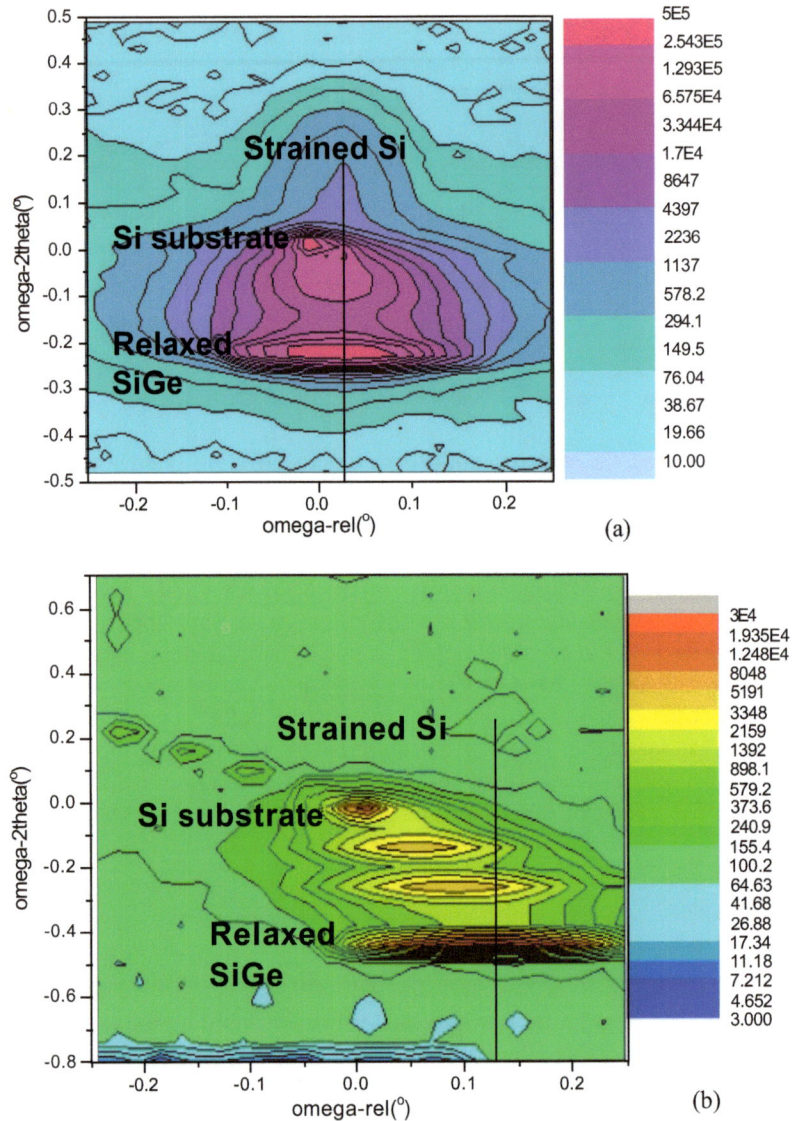

Figure 6: Contour plot of the (004) reciprocal space map of sample (a) S15 and (b) S25. Reprinted with permission 'Characterization of biaxial stressed silicon by spectroscopic ellipsometry and synchrotron X-ray scattering', T.K.S. Wong, Y. Gong, P. Yang, C.M. Ng, Semicond. Sci, and Technol. 22, 1232-1239, November 2007, IOP Publishing. DOI: 10.1088/0268-1242/22/11/009.

6. Sin²ψ METHOD

The need for strain characterization is not limited to the front end of the line or devices in an IC. The reliability of state of the art copper interconnects between devices is of great concern to the semiconductor industry because these interconnects like the gate electrodes are becoming narrower and narrower. As a result, the current density in these interconnects are increasing rapidly and this can result in electromigration (EM) failure [12]. A related interconnect reliability issue is stress migration (SM) which can take place in the absence of any current flow and is due to the presence of residual stress in the interconnect structure after fabrication. Since the metallization making up interconnects are either polycrystalline Damascence copper or aluminum, a different set of diffraction techniques are used for these materials and these will be discussed next.

The sin²ψ method is the oldest strain characterization method for polycrystalline materials based on XRD. It had been known since the 1920s and is discussed in detail in the book by Noyan and Cohen [13]. The basic approach of this method is to measure the inter-planar spacing of several related crystallographic planes that are inclined at different angles with respect to the sample surface (Fig. **7**). (The inclination angle ψ should not be confused with the angle φ to be defined in the next section.) The angle ψ is defined as the angle between the normal of a crystallographic plane and the sample surface normal (Fig. **7**) [21]. Thus if ψ = 0, the crystallographic plane is parallel to the sample surface.

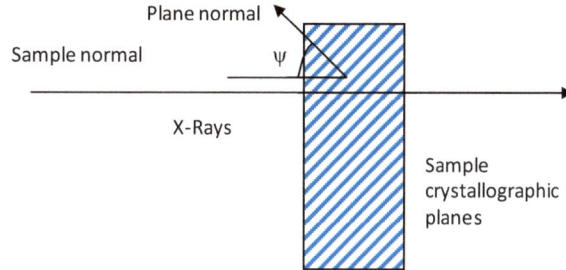

Figure 7: Principle of the $\sin^2 \psi$ method for polycrystalline samples.

We can consider an arbitrary crystallographic plane with lattice spacing d_0 in the absence of any stress. When stress is applied, this lattice spacing will change to a different value d. The strain in the direction perpendicular to the crystallographic plane is given by [13]:

$$\frac{d - d_0}{d_0} = \varepsilon_{11} \cos^2 \varphi \sin^2 \psi + \varepsilon_{12} \sin 2\varphi \sin^2 \psi + \varepsilon_{22} \sin^2 \varphi \sin^2 \psi + \varepsilon_{33} \cos^2 \psi$$
$$+ \varepsilon_{13} \cos \varphi \sin 2\psi + \varepsilon_{23} \sin \varphi \sin 2\psi$$

(11.4)

In this equation, the left hand side is simply due to the definition of strain. ε_{ij} are the strain tensor components referred to the coordinate system of the sample. Note that there are two coordinate systems involved here. The angle ψ as defined above is also the inclination between the sample coordinate system and the laboratory coordinate system. This latter coordinate system is referred to the diffracting planes. The surface normal of this plane is one of the axes of the laboratory coordinate system and it makes the angle ψ with respect to the sample surface normal. The second axis of the laboratory coordinate system is in the plane of the sample surface and it is at an angle ϕ with respect to the corresponding axis in the sample coordinate system.

By using the generalized Hooke's law, one can rewrite equation (11.4) in terms of the stress tensor components and the Young's modulus and Poisson's ratio of the material. For an isotropic material the equation (11.4) becomes [13]:

$$\frac{d - d_0}{d_0} = \frac{1 + v}{E} \Big[\sigma_{11} \cos^2 \varphi + \sigma_{12} \sin 2\varphi + \sigma_{22} \sin^2 \varphi - \sigma_{33} \Big] \sin^2 \psi +$$
$$+ \frac{1 + v}{E} \sigma_{33} - \frac{v}{E} \big[\sigma_{11} + \sigma_{22} + \sigma_{33} \big] + \frac{1 + v}{E} \big[\sigma_{13} \cos \varphi + \sigma_{23} \sin \varphi \big] \sin 2\psi$$

(11.5)

This is the generalized equation for the $\sin^2 \psi$ method. It can be considerably simplified for specific stress conditions. For example, for the case of simple biaxial stress which is often found in thin film systems, the stress σ_{33} (defined here as normal stress perpendicular to the sample) is zero and equation (11.5) becomes:

$$\frac{d - d_0}{d_0} = \frac{1 + v}{E} \Big[\sigma_{11} \cos^2 \varphi + \sigma_{22} \sin^2 \varphi \Big] \sin^2 \psi - \frac{v}{E} \big[\sigma_{11} + \sigma_{22} \big]$$

(11.6)

For the special case of eqi-biaxial stress σ_b, trigonometry allows further simplification to:

$$\frac{d-d_0}{d_0} = \frac{1+v}{E} \sigma_b \sin^2 \psi - \frac{2v}{E} \sigma_b \qquad (11.7)$$

This is because the in-plane stress components, $\sigma_{11} = \sigma_{22} = \sigma_b$. By looking at either equation (11.6) or (11.7), it is obvious that the inter-planar spacing d is a linear function of $\sin^2\psi$ for an isotropic material and for a given biaxial stress. This suggests the procedure of the $\sin^2\psi$ method which is as follows. First, select a family of crystallographic planes with Miller indexes $\{hkl\}$ and use XRD to measure the inter-planar spacing for several (at least two) crystallographic planes inclined at different angles ψ with respect to the sample surface. By plotting d *versus* $\sin^2\psi$, a straight line should be obtained. However, in order to deduce the strain, one obviously needs to know the equilibrium lattice spacing d_0. There are several ways to find this quantity. One approximation is to argue that there is no stress in the direction normal to the plane of the sample and therefore by extrapolating the plot of d *versus* $\sin^2\psi$ to $\psi = 0$, one can obtain d_0. From this, the strain for any ψ can be calculated.

One limitation of the $\sin^2\psi$ method is that diffraction has to be performed for several sets of related crystallographic planes. This can result in long measurement time for laboratory electron bombardment sources. Thus, in practice, it is common though not essential to carry out the $\sin^2\psi$ measurement at a synchrotron source.

6.1. Application Example

A simple example illustrating the $\sin^2\psi$ method is the study by Kraft *et al.* at the MaX-Planck Institute for metal research in Stuttgart [14]. The samples studied were 720 nm thick aluminum films deposited on a silicon substrate. The aluminum grains in these textured films are predominantly oriented in the (111) direction. As a result, the choice of inclination angle ψ is rather limited and only two $\sin^2\psi$ values were used for each measurement. In their data analysis, a linear relationship was assumed for the inter-planar spacing and $\sin^2\psi$. By drawing a line through the data points for each measurement and extrapolating, the value of d at $\sin^2\psi = 0$ and $\sin^2\psi = 1$ were found. These correspond respectively to the out of plane inter-planar spacing d_z and d_x respectively. As stated above, the equilibrium spacing is also needed in order to calculate the strain. The equilibrium spacing, d_0 is found from the angle ψ_0 for which the inter-planar spacing is not dependent on film stress. This angle is given by the equation [14]:

$$\sin^2\psi_0 = \frac{2C'_{13}}{2C'_{13} + C'_{33}} \qquad (11.8)$$

Here C'_{13} and C'_{33} are the stiffness constants of aluminum expressed in the sample coordinate system. This angle can also be found experimentally by measuring the d *versus* $\sin^2\psi$ data while heating up and cooling down the sample. The two lines should intersect at a point giving $\sin^2\psi_0$.

7. $\mathrm{Sin}^2\varphi$ METHOD

The $\sin^2\varphi$ method is an alternative strain characterization technique based on Laue diffraction of synchrotron X-rays in a transmission geometry. As will be stated below, the angle φ in the name of this method refers to the azimuth angle in the plane of the X-ray detector. This method was presented by Wanner and Dunand at the Materials Research Society Symposium in 2000 [15].

In this method, a monochromatic X-ray beam from a synchrotron source is incident normally at a bulk sample which is fabricated like a tensile test specimen (Fig. **8**). This is because tensile stress is applied *in situ* to the specimen during the test to induce a distortion of the diffraction pattern recorded by a charge coupled detector (CCD). In [15], a bulk Cu/Mo composite sample made by hot isotatic pressing was studied. When this sample is diffracted by the synchrotron X-ray beam, a series of diffraction rings are observed. These rings are circular for an unstressed sample and for small diffraction angles, the diameter D of the circular ring can be used to calculate the lattice spacing d by the Bragg equation:

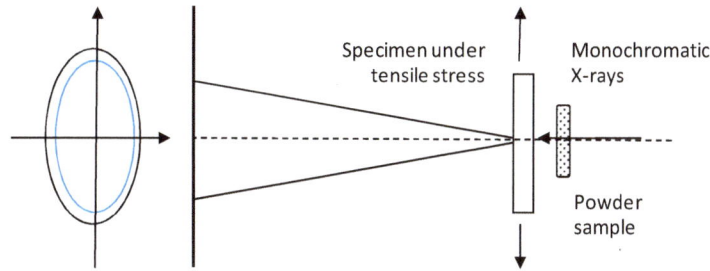

Figure 8: Principle of the $\sin^2\varphi$ for polycrystalline samples.

$$d \approx \frac{2\lambda L}{D} \tag{11.9}$$

Here, λ and L are the X-ray wavelength and the distance from the plane of the CCD to the sample. When uniaxial stress is applied to the specimen, D will change as a result. By using the above equation, the uniaxial strain can (in principle) be easily found. However, since λL may change during the course of the experiment, the following more robust formula is used to find the specimen strain ε_s [15]:

$$\varepsilon_s = \frac{d_s - d_s^0}{d_s} \approx 1 - \frac{D_s}{D_s^0} \cdot \frac{D_c^0}{D_c} \tag{11.10}$$

In the above, the subscripts s and c stand for specimen and calibration substance respectively and the superscript 0 represents a reference state. Thus, when this formula is applied, a second unstressed calibration substance needs to be placed in front of the specimen. It is also necessary that $L_c - L_s$ be constant and much smaller than L_s. For samples that give 'grainy' diffraction rings, the diameter can be measured as follows. For a given ring, the pixel intensity at the CCD is treated as a function of the radius R and the azimuth angle φ. (The geometric centre of the ring is the origin of the polar coordinate system.) For each angle φ, the pixel intensity as a function R is fitted to a Gaussian function and the peak of this function is used to obtain the radius at a particular φ. The radius values thus obtained for diametrically opposite points are added to give the diameter as a function of the azimuth.

During uniaxial tension tests, the circular rings will be distorted into ellipses. The vertical and horizontal diameters of a fitted ellipse are found from a plot of D(φ) versus $\sin^2\varphi$. A straight line is fitted to this plot and the intercept at $\sin^2\varphi = 1$ and $\sin^2\varphi = 0$ are used to find the diameter of the vertical (loading) and horizontal (transverse) directions respectively. These diameters are then substituted into equation (11.10) to yield the strain in these directions.

8. SCANNING X-RAY MICRODIFFRACTION

Both the diffraction methods described in the two previous sections have limited spatial resolution because of the X-ray beam size. In order to attain submicron spatial resolution and to perform mapping of the mechanical properties of a polycrystalline sample, the method of scanning X-ray microdiffraction (μSXRD) had been developed at two synchrotron facilities. These are the Advanced Photon Source (APS) [16] in Illinois and the Advanced Light Source (ALS) at the Lawrence Berkeley National Laboratory, University of California [17]. The key differences between μSXRD and the $\sin^2\psi$ and $\sin^2\varphi$ methods are that first, the sample is stationary in μSXRD and it is not tilted throughout the measurement. Second, the μSXRD method uses a polychromatic or white X-rays or a combination of monochromatic and polychromatic X-rays. It is important to understand these differences. Due to the mechanical limitations of present goniometers, the scattering volume in the sample will be changed as soon as the sample is tilted or rotated in the case of a small probe beam. In other words, it is not feasible to carry out a diffraction experiment with a monochromatic beam and rotate the sample and detector. For small volumes, one can only carry out diffraction in the alternative von Laue geometry in which multiple wavelengths are incident on a stationary sample at the same time. This generates a large number of diffracted beams and can be analyzed for strain information.

8.1. Beamline and End Station Arrangement

In the ALS implementation of the μSXRD technique [17], an X-ray beam with a photon flux rate of 10^6-10^7 $s^{-1}\mu m^{-2}$ is generated by a bending magnet synchrotron source. The X-ray beam at grazing incidence is first re-focused by a platinum coated toroidal mirror to the same size at the entrance of a hutch. The hutch is the enclosure within which the diffraction experiment took place. Refocusing is necessary because the incoming beam has a small but finite divergence in both the horizontal and vertical directions and there is a distance of 16m between the source and the mirror.

Inside the hutch are housed four main hardware components: (i) Si crystal monochromator, (ii) Kirkpartick-Baez (KB) mirrors, (iii) sample stage and (iv) CCD detector. The Si (111) monochromator used the Beaumont and Hart design [17] and has the property that the outgoing monochromatic beam can propagate in the same direction as the incident polychromatic (white) X-ray beam. The reason for using this configuration is that in experiments where a polychromatic beam is sufficient, the monochromator can simply be rotated out of the way and the white beam can propagate directly to the KB mirrors. If a monochromatic beam is needed, the monochromator can simply be rotated back in and the sample does not need to be moved.

The KB mirrors are able to focus a range of wavelengths to the same focal point. This is termed achromaticity in the literature [18]. In addition, the KB mirrors have high efficiency and can yield a small focal point. In order to produce the desired submicron focused beam, two orthogonal KB mirrors are needed, one for each direction. Each KB mirror is fabricated from a flat platinum coated piece fused silica. The flat mirror is bent into a precise elliptical shape by applying asymmetric couples using weak-leaf springs [18].

The sample is placed on two mechanical stages: a coarse Huber stage and a second precision piezoelectric stage. Note that sample translation only occurs between diffraction experiments. In other words, these stages allow a scanning microdiffraction experiment to be performed. A large CCD detector is used to collect the Laue diffraction pattern. The location and orientation of this detector should ensure a large number of diffracted beams to be recorded simultaneously.

8.2. Diffraction Image Analysis

By using the X-ray beamline described above, Laue diffraction patterns from a polycrystalline sample such as an aluminum film were first obtained. This pattern is obtained without moving the sample and it is the only result that is required for strain analysis. The diffraction pattern is analyzed by an image analysis software called X-MAS (X-ray microdiffraction analysis software) for the strain tensor. The key steps involved are highlighted in the following [17].

First, some preliminary image correction is performed on the CCD image of the Laue diffraction pattern. Then a two dimensional local maxima search routine is used to find the positions of the intensity peaks within the diffraction image. The accuracy of the strain analysis depends quite critically on the positions of these diffraction peaks. Hence, each located peak is fitted mathematically using a two dimensional Gaussian, Lorentzian or Pearson VII function to give sub-pixel resolution [17].

The next step is to find the correct Miller indexes of the peaks (Fig. **9**). The index of a peak can be found if the position of the CCD relative to the point of illumination on the sample and the incoming beam is known. This is the reason why a separate instrument geometry calibration involving an unstressed 'known' crystal such as silicon needs to be performed prior to indexing [17]. Suppose this step had been performed. From the wave vector of the incident beam k_{in} and the position of a peak in the plane of the CCD, the wave vector of the out-going diffracted beam k_{out} can be determined. Assuming elastic scattering, the experimental scattering vector q_{exp} is taken to be the vector difference between k_{out} and k_{in}. In other words, each peak in the CCD image corresponds to one experimental scattering vector. By iterating through the peaks of the image, a list of such scattering vectors can be obtained. From this list, a list of angles between the q_{exp} vectors is generated. This list of angles is then compared with a reference list of angles obtained from theoretical scattering vectors q_{ref} that are computed a structural model of the sample and for the energy range of the X-ray beam.

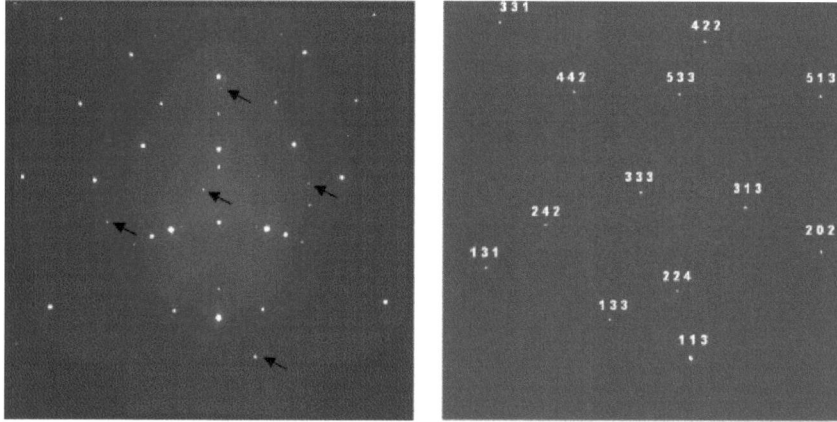

Figure 9: (left) Recorded Laue pattern of single grain aluminum interconnect on silicon. The pattern consists of both the diffraction spots from aluminum and silicon. (right) Laue pattern of aluminum after subtraction of silicon spots. Reprinted from *Nucl. Instru. Meth. Phys. Res. A*, vol. 467-468, A.A. MacDowell, R.S. Celestre, N. Tamura, R. Spolenak, B. Valek, W.L. Brown, J.C. Bravman, H.A. Padmore, B.W. Batterman and J.R. Patel, "Submicron X-ray diffraction", pp. 936-843 Copyright 2001 with permission from Elsevier.

In the X-MAS code reported [7], three scattering vectors are used and this subset is called a triplet. The X-MAS code looks for the best angular matches between a triplet from an experimental list and a triplet from a reference list subject to various constraints. When a match is found, a complete set of diffraction directions are calculated from the model and their location on the CCD are obtained using the geometry calibration. The best match is the one that can index correctly the largest number of experimental diffraction peaks.

Once the diffraction peaks are indexed, the crystal structure is basically known. The last step involves refining the values of the unit cell parameters of the crystal to obtain the deviatoric strain. There are six unit cell parameters: a, b, c, α, β, and γ. These are considered adjustable parameters while the other geometric such as the position of the illuminated point relative to the CCD are now fixed. By varying the unit cell parameters, one can again calculate the theoretical scattering vectors and the angles between them using the X-MAS code. These are compared and optimized with respect to the corresponding angles actually observed by the following minimization function:

$$\alpha_0 = \sum_i w_i \left(\alpha_i^{th} - \alpha_i^{exp} \right)^2 / \sum_i w_i \tag{11.11}$$

Here w_i are weighting factors and α_i^{th}, a_i^{exp} are the theoretical and experimental angle between two scattering vectors. The summation is over all pairs of α_i^{th}, a_i^{exp}. Once the unit cell parameters of the distorted cell are refined in this way, the transformation matrix t_{ij} which maps the lattice vectors of the undistorted to the distorted unit cell can be found. The deviatoric strain tensor is then found by using:

$$\varepsilon_{ij}' = \frac{\left(t_{ij} + t_{ji} \right)}{2} - I_{ij} \tag{11.12}$$

In equation (11.12), I_{ij} represents the identity matrix.

8.3. Application Example

The main application of the μSXRD technique thus far in microelectronics is in the EM characterization of metal interconnects. The grain size of metal grains in Al and Cu interconnects are typically in the range of 1μm or less. As a result, a conventional X-ray diffractometer does not have the mechanical precision to maintain incidence at the same region of the sample. In addition, the size of the X-ray probe from laboratory sources is much greater than the grain size. These limitations are overcome with the μSXRD technique.

In one study on EM in single damascene pure Cu interconnects [19], the Cu test structures were 1.1 μm wide, 100 μm long conductors with a thickness of 1 μm. The Cu surface was passivated by a layer of boron nitride. After EM testing at high current density and elevated temperatures, the same region at the anode end of the Cu interconnects were examined by both high voltage scanning electron microscopy (HVSEM) and μSXRD. HVSEM revealed the formation of a hillock near the anode which is the contact at which the current leaves the sample. The crystallographic orientation of the Cu grains and the shear stress in the same section of interconnect were found by μSXRD. The shear stress was calculated from the strain by using the stiffness tensor. For this experiment, the sample had to be scanned after each diffraction relative to the X-ray beam in order to generate the strain map. A sharp increase of stress was found at the location of the hillock consistent with HVSEM observation and could be attributed to the much lower surface diffusivity of Cu at the (115) twin compared with the (111) grains [19]. It is pointed out that in Cu EM, the main diffusion pathway is the interface between the metallization and the passivating dielectric. More recent work has been carried out to study plastic deformation in Cu interconnects by this method [20]

9. LIMITATIONS OF μSXRD

The spatial resolution of the μSXRD technique is at present limited by two factors: (i) X-ray probe spot size and (ii) scanning step size. The size of the spot at the sample is determined by the KB mirrors. Since the reported spot size is only in the submicron range [19], it is difficult to apply this technique to the gate electrode of state of the art semiconductor devices. When probe size in the deep submicron range can be obtained, the limiting factor will be the precision of the mechanical stages that translates the sample relative to the beam. Despite these limitations, the synchrotron scanning μSXRD method is a very useful EM characterization technique for metal interconnects.

REFERENCES

[1] D. K. Bowen and B.K. Tanner, *X-ray metrology in semiconductor manufacturing.* Taylor and Francis: Boca Raton, 2006.

[2] R. Jenkins, *X-ray fluorescence spectrometry.* Wiley: New York, 1999.

[3] J. Daillant and A. Gibaud, *X-ray and neutron reflectivity: principles and applications.* Springer Verlag: Berlin Heidelberg, 1999.

[4] Z. Bao, *Organic field-effect transistors.* CRC Press: Roca Baton, 2007.

[5] R.J. Matyi, "High resolution X-ray scattering methods for ULSI materials characterization," In: *Characterization and metrology for ULSI Technology*, 2005.

[6] T.K. Goh and T.K.S. Wong, "Investigation of thermal and oxygen plasma stability of mesoporous methylsilsesquioxane lowk films by X-ray reflectivity and small angle scattering," *Microelechon. Eng.*, vol. 73, pp. 330-343, Sep. 2004.

[7] B.D. Cullity, *Elements of X-ray diffraction.* Prentice Hall: Upper Saddle River, 2001.

[8] C. Riekel, "New avenues in X-ray microbeam experiments," *Rep. Prog. Phys.*, vol. 63, pp. 233-262, Mar. 2000.

[9] T.K.S. Wong, Y. Gong, P. Yang and C.M. Ng, "Characterization of biaxial stressed silicon by spectroscopic ellipsometry and synchrotron X-ray scattering," *Semicond. Sci. Technol.*, vol. 22, pp. 1232-1239, Nov. 2007.

[10] Z. Cai, "Studies of structural responses and stressor effects at the interfaces of Si/SiGe nanostructures using synchrotron X-ray nanodiffraction," In: *International Conference on Materials for Advanced Technologies abstracts*, 2007, pp. 9.

[11]   J. Zhang, X.B. Li and P.F. Fewster, "Dissecting a compositionally graded SiGe virtual substrate by X-ray reciprocal space mapping," In: *Electrochemical Society Proceedings volume 2004-07*, 2004, pp. 983-990.

[12]   B.C. Valek, J.C. Bravman, N. Tamura, A.A. MacDowell, R.S. Celestre, H.A. Padmore, B.W. Batterman and J.R. Patel, "Electromigration-induced plastic deformation in passivated metal lines," *Appl. Phys. Lett.,* vol. 81, pp. 4168-4170, Nov. 2002.

[13]   I.C. Noyan and J.B. Cohen, *Residual stress measurement by diffraction and interpretation.* Springer Verlag: New York, 1987.

[14]   O. Kraft, M. Hommel and E. Artz, "X-ray diffraction as a tool to study the mechanical behavior of thin films," *Mater. Sci. Eng. A,* vol. 288, pp. 209-216, Sep. 2000.

[15]   A. Wanner and D.C. Dunand, "Synchrotron X-ray study of elastic phase starins in the bulk of externally loaded Cu/Mo composite," In: *Materials Research Society Symposium Proceedings Vol. 590*, 2000, pp. 157-162.

[16] J.S. Chung, N. Tamura, G.E. Ice, B.C. Larson and J.D. Budai, "X-ray microbeam measurement of local texture and strain in metals," In: *Materials Research Society Symposium Proceedings Vol. 563*, 1999, pp. 169-174.

[17]   N. Tamura, A.A. MacDowell, R. Spolenak, B.C. Valek, J.C. Bravman, W.L. Brown, R.S. Celestre, H.A. Padmore, B.W. Batterman and J.R. Patel, " Scanning X-ray microdiffraction with submicrometer white beam for strain/stress and orientation mapping in thin films," *J. Synchrotron Rad.,* vol. 10, pp. 137-143, Mar. 2003.

[18]   A.A. MacDowell, R.S. Celestre, N. Tamura, R. Spolenak, B. Valek, W.L. Brown, J.C. Bravman, H.A. Padmore, B.W. Batterman and J.R. Patel, "Submicron X-ray diffraction," *Nucl. Instru. Meth. Phys. Res. A,* vol. 467-468, pp. 936-843, Jul. 2001.

[19] N. Tamura, R.S. Celestre, A.A. MacDowell, H.A. Padmore, R. Spolenak, B.C. Valek, N. Meier Chang and J.R. Patel, "Submicron X-ray diffraction and its application to problems in materials and environmental science," *Rev. Sci. Instru.,* vol. 73, pp. 1369-1372, Mar. 2002.

[20]   A.S. Budiman, W.D. Nix, N. Tamura, B.C. Valek, K. Gadre, J. Maiz and J.R. Patel, "Crystal plascticity in Cu damascene interconnect lines undergoing electromigration as revealed by synchrotron X-ray microdiffraction," *Appl. Phys. Lett.,* vol. 88, pp. 233515-1-3, Jun. 2006.

[21]   R. Spolenak, W. Ludwig, J.Y. Buffiere and J. Michler, "*In situ* elastic strain measurements- diffraction and spectroscopy," *MRS Bull.,* vol. 35, pp. 368-374, May. 2010.

CHAPTER 12

Conclusion and Outlook

Abstract: Some general remarks are given for all the strain characterization methods discussed in previous chapters. The methods can be categorized into those where a strain related material property must be known and those that do not require such a priori knowledge. The different available methods are briefly compared with each other in terms of the strain resolution, spatial resolution, throughput and accessibility. The future application of semiconductor strain characterization techniques in electrical energy storage devices is suggested and a recent development of strain characterization by electron backscattering from epitaxial layers is highlighted.

In this volume, we have endeavoured to provide an up to date and cross disciplinary survey of current strain characterization methods for semiconductor thin films, nanoelectronic and microelectromechanical devices. Some earlier methods such as the ion channeling method had been left out in this discussion because they had been discussed in previous publications or reviews [1]. There is clearly a wide variety of techniques for strain characterization in semiconductors. By referring to the explanation of each technqiue in the previous chapters, a judicious choice on the most appropriate method for a strain measurement problem can be made. In applying a given method, the limitations as well as the capabilities of each technique should be carefully considered.

The many different methods for characterizing strain in semiconductors can be divided into two groups. The first group is based on the fact that strain changes the structure and hence symmetry of a semiconductor crystal. When the lattice parameters of the unit cell are changed, the electronic band structure of the crystal will be modified because of a change in crystal symmetry. Degenerate electronic states in the unstrained crystal can become distinct in energy and the curvature of the energy bands (or carrier effective mass) will become different from that of the unstrained crystal. Similarly, the energy spectrum (or dispersion relation) for phonons is changed by an applied strain. The modification of the electronic structure and phonon spectrum with strain is typically linear for small strains. Thus, by monitoring the changes in the optical or electrical properties of a strained sample, the value of some components of the strain tensor can be indirectly determined. Optical properties that had been used to deduce strain include the complex refractive index, the luminescence spectrum and the Raman spectrum of a semiconductor. For electrical property, the mobility or resistivity of the sample is often used. For all the indirect methods, certain material properties such as the piezospectrosopic coefficient need to be known from prior calibration experiments. This requirement may present a difficulty for new materials that are studied for the first time.

The second group of methods is based on the diffraction or holography of electrons and X-rays. These methods are direct in the sense that they involve measurement of the semiconductor lattice spacing. Since the normal component of strain is defined as a fractional change in length, both the unstrained and strained regions of the same sample must be measured. While these direct methods do not require prior knowledge of certain material specific properties, sample preparation can be extremely challenging. Thus, the availability of technical personnel with expertise in sample preparation may be the practical constraint for these techniques.

In the following, we briefly compare the present strain characterization techniques in terms of strain resolution, spatial resolution, throughput and accessability. The latter is seldom mentioned but is clearly of practical importance. First, both the TEM and X-ray methods have the highest overall strain resolution. The SE, Raman and CL methods have relatively lower strain resolution because of uncertainty over the strain related material parameter. In terms of spatial resolution, the TEM methods are again the best because of the small de-Broglie wavelength of high energy electrons. The TEM holographic method in particular is uniquely capable of both high spatial resolution and a large measurement area. The AFM-DIC method also has high spatial resolution because of the use of sharp scanning probes. Due to the difficulty of focusing X-rays, it is only gradually becoming possible to achieve high spatial resolution for X-ray diffraction methods. For all the optical methods (SE, PR, micro-Raman), the wavelength of light limits the spatial resolution to the micron scale. Application of near field nanophotonic techniques such as TERS, however, can extend the spatial resolution to the nanoscale. In future, TERS should become more and more widely used because of this. The third criterion for comparison is the throughput. This refers to the total amount of time needed to measure a strained semiconductor sample. For those techniques that involve sample preparation, the sample preparation time has to be included. On this basis, the TEM methods are considered low throughput because considerable effort is needed to prepare a usable TEM sample. For X-ray diffraction methods, long integration time is needed to achieve a good quality serial diffraction scan with present laboratory X-ray sources. X-ray diffraction can be considered as low throughput. However, this is subject to change in future. The fastest methods are the optical methods (SE, PR and micro-Raman) because they require no sample preparation. If the semiconductor being studied has known material properties such as silicon or germanium, the strain can be determined very quickly.

The fourth comparison concerns the accessability of the strain measurement. Although the TEM methods have high strain resolution, high spatial resolution and for holography, a large field of view, the methods are not readily available because they require expensive state of the art electron microscopes, quantitative electron microscopy software and highly trained personnel. X-ray microdiffraction methods require access to large synchrotron facilities where a prospective user has to submit a proposal for the experiment to be performed months in advance. The proposed experiment is then peer reviewed for final acceptance. Thus, despite the fact that there are numerous methods available, a high resolution strain measurement with high spatial resolution over a large sample area is still a challenge for most research groups.

As mentioned in chapter 1, the motivation for developing many of the strain characterization techniques in this book is due to the introduction of strained microelectronic devices by the semiconductor industry. However, the scope of application is not limited to this industry alone. In the future, one can anticipate growing applications of these techniques in photovoltaic devices and electrical energy storage devices such as rechargeable batteries. This is because in photovoltaics, strain can modify the absorption spectrum of the semiconductor and adversely affect the power conversion efficiency even for a well designed cell structure. For batteries, the strain from repeated charging and discharging can weaken the electrode material over time and limit the charge discharge cycle life of the battery.

In one recent study by Harris and co-workers at the General Motors R&D center [2], the fracture strength of $LiCoO_2$ and graphite electrode particles of lithium ion batteries were found to depend on their internal microstructure such as pores and cracks. Particle fracture was often observed in degraded battery electrodes and internal surfaces modified the stress fields. However, at the moment, these are not well understood. The measurement of the internal stress fields by the methods discussed in this volume may help to improve understanding on this important problem for future electric transportation.

Finally, at the conclusion of this writing, another new strain characterization method based on electron backscattering has been reported by Tomita *et al.* at Meiji University, Japan [3]. This method is essentially electron diffraction. However, the measurement is performed inside a SEM because only the backscattered electrons are collected. As a result, sample thinning as in CBED is not required and stress relaxation effects can be eliminated. The investigators have demonstrated strain tensor measurements in SSOI substrates and in silicon wafers with patterned silicon nitride. This development is testimony to the dynamic nature of this field in semiconductor metrology.

REFERENCES

[1]   H.J. Grossman, B.A. Davidson, G.J. Gualtieri, G.P. Schwartz, A.T. Macrander, S.E. Slusky, M.H. Grabow and W.A. Sunder, "Critical layer thickness and strain relaxation measurements in GaSb(001)/AlSb structures," *J. Appl. Phys.*, vol. 66, pp. 1687-1693 Aug. 1989.

[2]   S.J. Harris, R.D. Deshpande, Y. Qi, I. Dutta and Y.T. Cheng, "Mesopores inside electrode particles can change Li-ion transport mechanism and diffusion-induced stress," *J. Mater. Res.*, vol. 25, pp. 1433-1440 Aug. 2010.

[3]   M. Tomita, D. Kosemura, M. Takei, K. Nagata, H. Akamatsu and A. Ougra, "Evaluation of strained-silicon by electron backscattering pattern measurement: comparison study with UV-Raman measurement and edge force model calculation," *Jap. J. Appl. Phys.*, vol. 50, pp. 010111-1-8, Mar. 2011.

APPENDIX 1

A. Energy Momentum (*E-k*) Diagram of Unstrained Bulk Silicon Computed Using StrainBands [1]

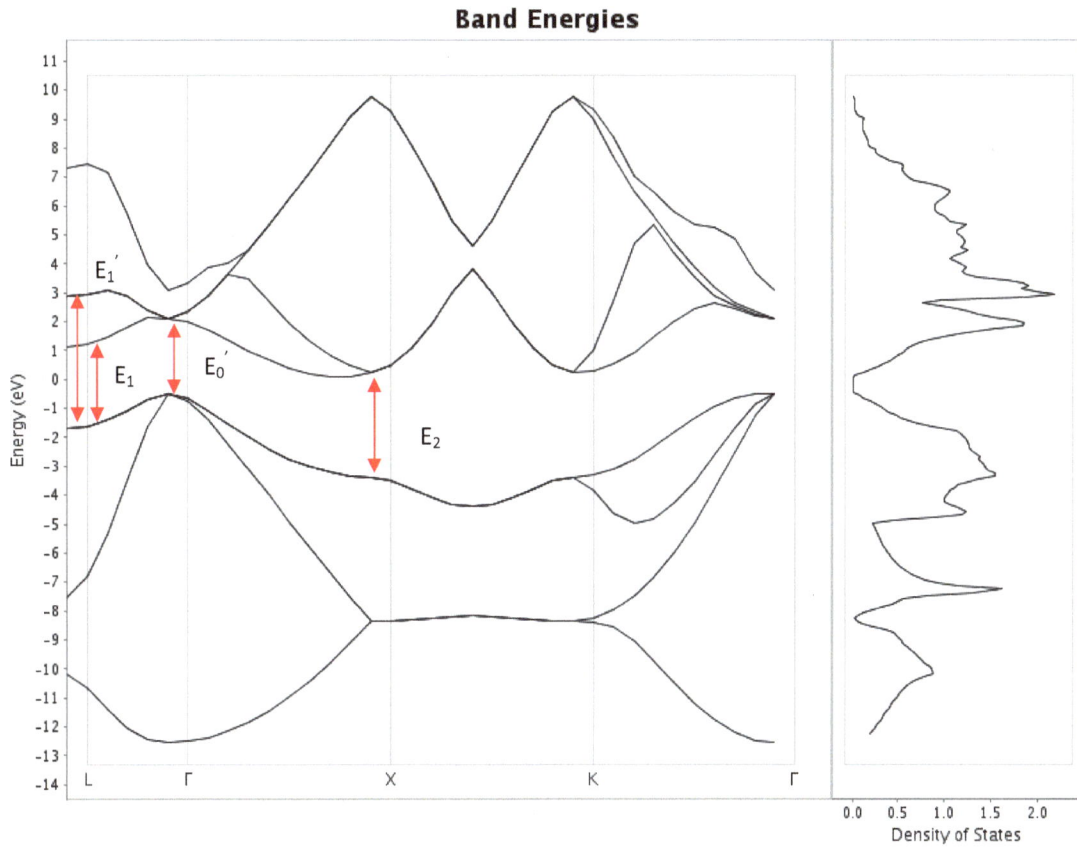

Band Energies

[1] J. Ringgenberg, J. Bhattacharjee, J.B. Neaton, J.C. Grossman, E. Schwegler (2008), "StrainBands" DOI: 10254/nanohub-r2815.2. (DOI: 10254/nanohub-r2815.2).

B. Computed Energy-Momentum Diagram of Biaxial Strained Silicon with 0.5% Strain with Respect to Bulk Silicon [1]

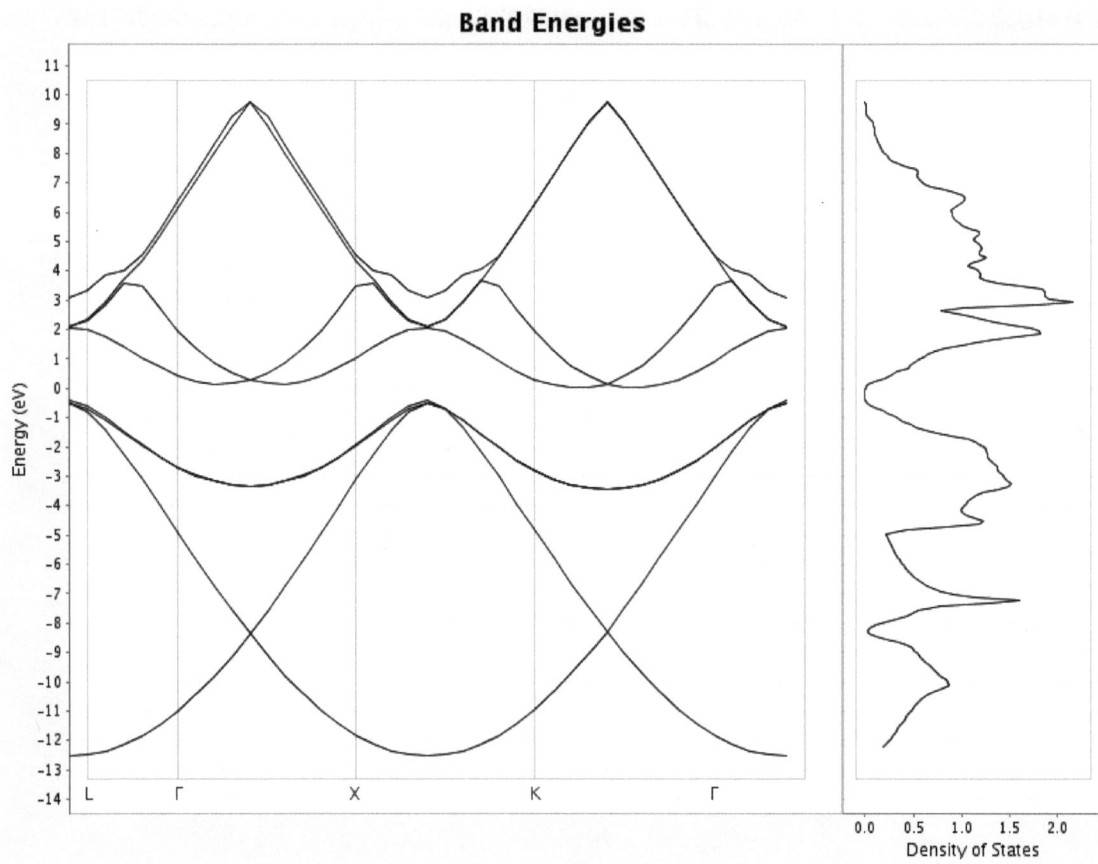

Band Energies

APPENDIX 2

Table A2.1: Stiffness and Compliance Constants for Selected Cubic Semiconductors at 300K [2]. The Unit for C_{ij} is 10^{10} Pa and the Unit for S_{ij} is 10^{-13} Pa.

	C_{11}	C_{12}	C_{44}	S_{11}	S_{12}	S_{44}
Si	16.564	6.394	7.951	0.7691	-0.2142	1.2577
Ge	12.87	4.77	6.67	0.9718	-0.2628	1.499
GaAs	11.88	5.38	5.94	1.173	-0.366	1.684
InP	10.22	5.73	4.42	1.639	-0.589	2.26
InAs	8.329	4.529	3.959	1.945	-0.6847	2.526
AlAs	11.93	5.72	5.72	1.216	-0.394	1.748
c-AlN	31.5	15.0	18.5	0.458	-0.148	0.541
CdTe	5.35	3.69	2.02	4.27	-1.74	4.95

Table A2.2: Young's modulus Y, Poisson Ratio P and Shear Modulus γ for Selected Cubic Semiconductors at 300K [2] The Unit for Y is 10^{11} Pa and the Unit for γ is 10^{10} Pa. m is the Longitudinal Stress Direction and n is the Transverse Direction.

	Y (100)		P (100)	γ
	[001]	[011]	m=[010]; n=[001]	
Si	1.3	1.69	0.279	5.085
Ge	1.029	1.371	0.270	4.05
GaAs	0.853	1.213	0.312	3.25
InP	0.610	0.917	0.359	2.25
InAs	0.514	0.793	0.352	1.90
AlAs	0.822	1.179	0.324	3.11
c-AlN	2.18	3.45	0.323	8.25
CdTe	0.234	0.400	0.407	0.830

[2] S. Adachi, Properties of Group-IV, III-V and II-VI semiconductors. Wiley: Chichester, 2005.

INDEX

www.ingramcontent.com/pod-product-compliance
Lightning Source LLC
Chambersburg PA
CBHW041715210326
41598CB00007B/655